STUDENT SOLUTIONS MANUAL FOR

INTERMEDIATE ALGEBRA
THIRD EDITION

Barker ▪ Rogers ▪ Van Dyke

Grace Malaney
Donnelly College

Saunders College Publishing
Harcourt Brace Jovanovich College Publishers
Fort Worth Philadelphia San Diego
New York Orlando Austin San Antonio
Toronto Montreal London Sydney Tokyo

Copyright © 1991, 1987 by Saunders College Publishing

All rights reserved. No part of this publication may be reproduced or transmitted in any form or by any means, electronic or mechanical, including photocopy, recording or any information storage and retrieval system, without permission in writing from the publisher.

Requests for permission to make copies of any part of the work should be mailed to: Permissions, Harcourt Brace Jovanovich, Inc., Orlando, Florida 32887.

Printed in the United States of America.

ISBN 0-03-052848-8

234567   095   9876543

PREFACE

This Student Solutions Manual is to accompany INTERMEDIATE ALGEBRA, third edition, by Barker, Rogers and Van Dyke. This manual includes solutions for every other odd-numbered problem of each exercise set. Instead of solutions only, word explanations accompany the symbolic solutions, which are often written out in considerable detail. It should not be necessary to ask, "Now what did she do?" In most cases, what is offered is a "no steps left out" solutions manual.

Note to Students:

To use this manual I would recommend that it be placed to one side of a pad of paper with the text on the other side. If you are successful in solving the problem then you may or may not refer to the manual. However, if after trying to solve the problem you are unable to solve it, or if you are not sure of your solution and would like to check it by comparing your work to the manual solution, or if you would just like to look at another solution, then refer to the manual and follow through the solution there.

The manual may also be used to study for examinations by working your way through the solutions (don't just read or glance at them) and the word explanations.

Note to Instructors:

This manual may also be used by instructors as an aid in selecting homework assignments and examination items. By glancing through the manual and looking at the length of the solutions, it is possible to get an idea of the difficulty of the problems.

I have worked hard trying to make the manual free of both typing errors and conceptual errors. If any errors have gone undetected, I would appreciate it very much if you would bring them to my attention. Thank you.

Grace Malaney
Mathematics and Science Division
Donnelly College
608 North 18th Street
Kansas City, Kansas 66102

TABLE OF CONTENTS

		Page
1.	Properties of Real Numbers	1
2.	Polynomials	31
3.	Rational Expressions	73
4.	Roots and Radicals	103
5.	Quadratic Equations	131
6.	Relations and Functions	193
7.	Systems of Linear Equations and Inequalities	235
8.	Second-Degree Equations in Two Variables	267
9.	Exponential and Logarithmic Functions	305
10.	Functions of Counting Numbers	335

CHAPTER 1

PROPERTIES OF REAL NUMBERS

EXERCISES 1.1 REAL NUMBERS: A REVIEW

A

Given: $-12, -8, -\frac{9}{2}, -0.32, 0, \frac{1}{2}, 1, 1.84, \frac{15}{7}, \sqrt{5}, \sqrt{8}, 47, 134.7$

1. List the whole numbers:

   $\{0, 1, 47\}$ There are three whole numbers.

5. List the positive real numbers.

   $\left\{\frac{1}{2}, 1, 1.84, \frac{15}{7}, \sqrt{5}, \sqrt{8}, 47, 134.7\right\}$ All the numbers in the given set are real numbers. Select the numbers that are positive.

Answer true or false:

9. $0 < 5$
   True.   0 is to the left of 5.

13. 2.3 is an integer.
    False.   Integers are either whole numbers or their opposites.

B

Answer true or false:

17. $-12, 7, 0, -34, 681$ are all integers.
    True.   The set of integers includes the set of negative integers and the set of whole numbers.

21. $14, 28, 75, 1821, 6429$ are all real numbers.
    True.   All positive whole numbers are real numbers.

25. -21 is a rational number.
    True.

    Any number that can be written as a fraction in which both numerator and denominator are integers and the denominator is not zero is a rational number.
    $\left[-21 \text{ can be written } -\frac{21}{1}.\right]$

True or False?

29. -3.221221... is a rational number.
    True.

    This is a nonterminating repeating decimal, and is therefore a rational number.

True or false?

33. -62 < 0
    True.

    -62 is to the left of 0.

C

Use -, <, or > in the box to make a true statement.

37. 5 $\boxed{<}$ 12

    5 is to the left of 12 so use <.

41. 0 $\boxed{<}$ 5

    0 is to the left of 5 so use <.

45. -8 $\boxed{<}$ -4

    -8 is to the left of -4 so use <.

True or False?

49. 8 ≠ 7
    True.

    The symbol "≠" indicates that 8 is not equal to 7.

53. -121 ≠ -121
    False.

    -121 equals -121.

EXERCISES 1.2   LAWS OF REAL NUMBERS

A

The following are examples of what laws of real numbers?

1. $-8 + 4 = 4 + (-8)$
   Commutative property of addition.

   The order of addends has been changed.

5. $-8 \cdot 1 = -8$
   Multiplication property of one.

   If a is a real number, than $a \cdot 1 = 1$.

Write the reciprocals of each of the following:

9. 12
   The reciprocal of 12 is $\frac{1}{12}$.

   $12 \cdot \frac{1}{12} = 1$

13. 1
    The reciprocal of 1 is 1.

    $1 \cdot 1 = 1$

B

Answer true or false:

17. $2 \cdot 3 + 4 = 2 \cdot 3 + 2 \cdot 4$ is an example of the Distributive property.
    False.

    In $2 \cdot 3 + 4$, the 2 is being multiplied times 3, not the sum $(3 + 4)$.

21. $7 + (-7) = 0$ is an example of the addition property of opposites.
    True.

    7 and -7 are opposites.

25. $3(x + y) = (3x) + y$ is an example of the associative property of addition.
    False.

    $3(x + y) \neq (3x) + y$
    A false statement cannot be an example of any of the laws of real numbers.

The following is an example of what law of real numbers?

29. $(7 + 9)4 = 7 \cdot 4 + 9 \cdot 4$
    Distributive property.

    This law changes a product to a sum.

3

33. $\left[\dfrac{2}{x}\right]\left[\dfrac{x}{2}\right] = 1$

   Multiplication property of reciprocals.   The product of reciprocals is 1.

C

Justify each step in the solution of each of the following problems:

37. $3 + 4(-8 + 8) = 3 + 4(0)$
    Addition property of opposites $-8 + 8 = 0$

41. $3 + 1(-3) + 1(8) = 3 + (-3) + 8$
    Multiplication property of one $1(-3) = -3$ and $1(8) = 8$

45. $-12 + 4(3 + 2) = -12 + 12 + 8$
    Distributive property $4(3 + 2) = 4\cdot 3 + 4\cdot 2 = 12 + 8$

49. $-1(2x + 2y) = -2x + (-2y)$
    Distributive property $-1(2x + 2y) = (-1)2x + (-1)2y$

53. $5 + 2(3 + 2(1 + (-1))) = 5 + 2(3 + 2(0))$
    Addition property of opposites $1 + (-1) = 0$

57. $5 + 6 = 11$ Closure property of addition.

EXERCISES 1.3 OPPOSITES, ABSOLUTE VALUE, ADDITION OF REAL NUMBERS

A

Perform the indicated operations:

1. $|-6| = 6$ The number -6 is 6 units from zero on the number line.

5. $-|-4|$
   $= -4$ "The opposite of the absolute value of -4."
   First find the absolute value of -4. Then find the opposite.

9. $-8 + (-2)$
   $= -10$ The sum of two negative numbers is negative.

13. $-(4 + 1)$
    $= -5$ Add $4 + 1$. Find the opposite.

B

17. $|8 + 7|$
    $= |15|$
    $= 15$

    Find the sum of 8 + 7. Then find the absolute value of the sum. The distance between 15 and 0 is 15.

21. $-(5 + 12)$
    $= -(17)$
    $= -17$

    Find the sum. 5 + 12 = 17. The opposite of 17 is -17.

25. $-8 + (-3) + 5$
    $= -11 + 5$
    $= -6$

    Add in order from left to right.

29. $-6 + (-8) + (-12)$
    $= -14 + (-12)$
    $= -26$

    Add in order, left to right.

33. $-[x + (-2)]$
    $= -x + 2$ or $2 - x$

    To find the opposite of a quantity, find the opposite of each member of the quantity.

C

37. $-(15 + 12) + (-36)$
    $= (-15) + (-12) + (-36)$

    $= (-27) + (-36)$
    $= -63$

    Find the opposite of 15 + 12 by writing the opposite of 15 and the opposite of 12.
    Add in order from left to right.

41. $-(3.2) + (-7.5) + (1.9)$
    $= (-3.2) + (-7.5) + (1.9)$
    $= (-10.7) + (1.9)$
    $= -8.8$

    $-(3.2) = -3.2$

45. $-[-|-4| + (-|3|)] + [3 + (-2)]$
    $= -[(-4) + (-3)] + [3 + (-2)]$

    $= 4 + 3 + 3 + (-2)$

    $= 7 + 3 + (-2)$
    $= 10 + (-2)$
    $= 8$

    The absolute value of -4 is 4, and its opposite is -4. The absolute value of 3 is 3, and its opposite is -3. The opposite of -4 is 4, and the opposite of -3 is 3.

49. $-|-4 + 7| + |-16 + 12| + |-12 + 2|$
    $= -|3| + |-4| + |-10|$
    $= (-3) + 4 + 10$
    $= 1 + 10$
    $= 11$

    Perform the operation inside the absolute value symbols. $-|3| = -3$, $|-4| = 4$, and $|-10| = 10$.

53. |5 + (-8)| + |-16 + (-4)| + (-|-5 + (-9)|)
   = |-3| + |-20| + (-|-14|)   Perform the operations inside each set of absolute values symbols.
   = 3 + 20 + [-(14)]   The absolute value of negative numbers is positive.
   = 3 + 20 + (-14)   The opposite of a postiive number is a negative number.
   = 23 + (-14)   Add in order from left to right.
   = 9

D

57. The checking account of the Square Hole Donut Company showed credits of $894.72 and debits of $674.72 (-$674.72) for one day. What increase or decrease will the balance of the account show?

    894.72 + (-674.72)   Find the sum.
       = 220

    Since the sum is positive, there was an increase of $220 in the account.

61. Jan Long has a checking account balance of $641.32. She wrote checks for $49.50 and $36.47, then made a deposit of $50.00. What is her balance?

    641.32 + (-49.50) + (-36.47) + 50.00
      = 591.82 + (-36.47) + 50.00   Add from left to right.
      = 555.35 + 50.00
      = 605.35

    Jane's current balance is $605.35.

65. Harry and Tom formed a partnership. They agree to split all earnings and expenses equally between each other. If Harry earned $172.50 and Tom earned $156.25, what is the net worth of the partnership if Harry's expenses were $33.15 and Tom's expenses were $28.35?

    172.50 + 156.25 + (-33.15) + (-28.35)   Write the earnings as positive numbers and the expenses as negative numbers.
                                             Find the sum.
      = 328.75 + (-33.15) + (-28.35)
      = 295.60 + (-28.35)
      = 267.25

    The net worth of the partnership is $267.25.

EXERCISES 1.4   SUBTRACTION OF REAL NUMBERS AND COMBINING TERMS

A

Perform the indicated operations (subtract or combine like terms):

1. $12 - 6$
    $= 12 + (-6)$
    $= 6$

    To subtract a number, add its opposite.

5. $-8 - (-4)$
    $= -8 + 4$
    $= -4$

    Rewrite as addition by adding the opposite.

9. $7a + 4a$
    $= (7 + 4)a = 11a$

    Use the distributive property to factor, then find the sum of the numerical coefficients.

13. $-8abc - 9abc$
    $= (-8 - 9)abc$
    $= -17abc$

    Use the distributive property to factor, then find the difference of the numerical coefficients.
    $-8 - 9 = (-8) + (-9) = -17$

B

17. $115 - 291$
    $= 115 + (-291)$
    $= -176$

    Rewrite as addition by adding the opposite.

21. $\frac{5}{8} - \frac{1}{4}$
    $= \frac{5}{8} - \frac{2}{8}$
    $= \frac{5}{8} + \left[-\frac{2}{8}\right] = \frac{5}{8} + \frac{-2}{8}$
    $= \frac{5 + (-2)}{8} = \frac{3}{8}$

    Write each fraction using the common denominator.
    Rewrite as addition by adding the opposite.

25. $-0.5 - (-0.4)$
    $= -0.5 + (0.4)$
    $= -0.1$

    Rewrite as addition by adding the opposite.

29. $6x + 5x - 7x$
    $= (6 + 5 - 7)x$

    $= 4x$

    Use the distributive property to factor; then find the sum of the numerical coefficients.

33. $6x + 8 + 12x$
    $= (6 + 12)x + 8$
    $= 18x + 8$

    Factor the like terms.

C

37. $-8.7 - 14.6$
    $= -8.7 + (-14.6)$
    $= -23.3$

To subtract, add the opposite. The sum of two negative numbers is negative.

41. $22.5 - (-16.3)$
    $= 22.5 + 16.3$
    $= 38.8$

Rewrite as addition by adding the opposite.

45. $4x + 7y - 8x - 4y$
    $= -4x + 3y$

Combine the like terms $4x$ and $-8x$ and the like terms $7y$ and $-4y$.

49. $-\frac{1}{3}a - \frac{1}{4}b - \frac{1}{12}a - \frac{1}{2}b$
    $= -\frac{1}{3}a - \frac{1}{12}a - \frac{1}{4}b - \frac{1}{2}b$
    $= -\frac{5}{12}a - \frac{3}{4}b$

Combine the like terms.

53. $3.52y - 12 + 6.3z + 14$
    $= 3.52y + 6.3z + 2$

Combine the like terms: $-12 + 14$.

D

57. What was the change in temperature (difference) from a low of $-8°F$ to a high of $40°F$?

    $40 - (-8)$
    $= 40 + 8$
    $= 48$

To find the difference, subtract.

There was a 48° change in temperature.

61. What was the change in altitude from 250 ft below sea level (-250) to 600 ft above sea level?

    $600 - (-250)$
    $= 600 + (250)$
    $= 850$

To find the change, or difference, subtract. Add the opposite.

The change in altitude was 850 ft.

65. If a check is written for the amount of $85.42 on an account with a balance of $96.57, what is the new balance?

    96.57 - (85.42)   Subtract the amount of the check from the balance.

    = 96.57 + (-85.42)
    = 11.15

The new balance is $11.15.

EXERCISES 1.5 MULTIPLICATION AND DIVISION OF REAL NUMBERS

A

Perform the indicated operations:

1. (-7)(-5)
    = 35   The product of two negative numbers is positive.

5. (-12)(0)
    = 0   The product of any number and zero is zero.

9. $-64 \div 8$
    = -8   The quotient of a negative number and a positive number is negative.

13. $(-10x) \div (-2)$
    $= \dfrac{-10x}{-2}$ Write the division as a fraction.
    = 5x   Divide the coefficient of x by -2. $(-10) \div (-2) = 5$.

B

17. (-2)(-3)(4)
    = (6)(4)   Multiply from left to right.
    = 24

21. $\dfrac{-50}{-5}$
    = 10   The quotient of two negative numbers is positive.

25. (-3a)(4)(-2)
    = (-12a)(-2)   Multiply the numbers from left to right.
    = 24a

29. $3x \div 5$
    $= \dfrac{3x}{5}$ or $\dfrac{3}{5}x$ Write the division as a fraction.

33. $-(3)^2(-2)(5)$
    $= (-9)(-2)(5)$
    $= (18)(5)$
    $= 90$

    The opposite of $(3)^2$ is -9.

C

37. $[(6x)(-5)] \div 2$
    $= (-30x) \div 2$

    $= \dfrac{-30x}{2} = -15x$

    Find the product within brackets.
    Then write the division as a fraction, and divide the coefficients.

41. $\dfrac{(-5z)(2)}{7}$

    $= \dfrac{-10z}{7}$ or $-\dfrac{10}{7}z$

    Find the product in the numerator.

45. $6(-2x + 4)$
    $= -12x + 24$

    Multiply using the distributive property.

49. $-5(-5a + 2b - 3)$
    $= 25a - 10b + 15$

    Distributive property.

53. $-8(-3a - 4b + 2)$
    $= 24a + 32b - 16$

    Distributive propertyy.

D

57. Use the formula $C = \dfrac{5}{9}(F - 32)$ to change $-13°F$ to a Celsius measure.

    $C = \dfrac{5}{9}(F - 32)$ Formula.

    $C = \dfrac{5}{9}(-13 - 32)$ Replace F with -13.

    $C = \dfrac{5}{9}(-45) = \dfrac{5}{\cancel{9}_1} \cdot \dfrac{\cancel{-45}^{-5}}{1}$ Simplify.

    $C = -25$

    So $-13°F$ equals $-25°C$.

61. In a business a cash flow can be either positive or negative. If the Fillum-Up service station had an average cash flow of -$20 a day for 10 days, what was the cash flow for that period?

    10(-20)
      = -200   The cash flow for the 10-day period can be represented by the product of 10 and -20.

The cash flow for the 10-day period was -$200, or a loss of $200.

65. A plane is descending at a rate of 150 ft every minute (-150). If it was originally 1020 ft above sea level, what is its elevation after five minutes?

    5(-150)   The change in elevation is represented by the product of 5 and -150.

    1020 + 5(-150)   The sum of the old elevation
      = 1020 + (-750)   and the change in elevation
      = 270   represents the new elevation.

Its elevation is 270 ft above sea level.

EXERCISES 1.6   ORDER OF OPERATIONS

A

Perform the indicated operations:

1. -5(-4 + 3)
    = -5(-1)   Do the addition inside the
    = 5   parentheses first. Then multiply.

5. $(2 + 3)^2$
    = $(5)^2$ Do the addition inside the
    = 25   parentheses first. Then find the square.

Simplify:

9. 4·3x - 2·5x
    = 12x - 10x   Do the two multiplications
    = 2x   first. Combine like terms.

Evaluate the following if x = -5:

13. x - 12
     -5 - 12
      = -5 + (-12) = -17   Replace x with -5 and simplify.

B

Perform the indicated operations:

17. $12 - 3(-8 + 4) - 7$
   $= 12 - 3(-4) - 7$ Do the addition inside the parentheses first.
   $= 12 + 12 - 7$ Multiply.
   $= 24 - 7$ Add and subtract from left to right.
   $= 17$

21. $(-2)^2(4) - 4(5)^2$
   $= (4)(4) - 4(25)$ Perform the operations indicated by the exponents.
   $= 16 - 100$ Multiply.
   $= -84$ Subtract.

Simplify:

25. $4a - 3b - (3b + 4a)2$
   $= 4a - 3b - 6b - 8a$ Use the Distributive property to clear the parentheses.
   $= -4a - 9b$ Combine like terms.

29. $3(-6x - 7) - 4(-2x + 6)$
   $= -18x - 21 + 8x - 24$ Use the Distributive property to clear both parentheses.
   $= -10x - 45$ Combine like terms.

Evaluate the following if $a = -2$ and $b = 2$:

33. $2a + 3b - 5(a + 1)$
   $2(-2) + 3(2) - 5(-2 + 1)$ Replace a with -2 and b with 2. Do the addition inside the parentheses first.
   $= 2(-2) + 3(2) - 5(-1)$
   $= -4 + 6 + 5$ Do the multiplication from left to right. Add from left to right.
   $= 7$

37. $4a - 3b - 2(a + b)$
   $4(-2) - 3(2) - 2(-2 + 2)$ Replace a with -2 and b with 2. Do the addition inside the parentheses first.
   $= 4(-2) - 3(2) - 2(0)$
   $= -8 - 6 - 0$ Multiply from left to right.
   $= -14$ Subtract.

C

Perform the indicated operations:

41. $-2(8 - 12) - 3[2(8 - 4)] - 3$
    $= -2(-4) - 3[2(4)] - 3$ Do the subtractions inside
    $= -2(-4) - 3[8] - 3$ the parentheses first.
    $= 8 - 24 - 3$ Multiply inside brackets.
                                 Multiply from left to right.
    $= -19$ Subtract from left to right.

45. $-5 + 3(-2)^2 - [(4 - 5)^2 - 8(-2 - 4)]$
    $= -5 + 3(-2)^2 - [(-1)^2 - 8(-6)]$ Do the subtractions inside
                                              parentheses.
    $= -5 + 3(4) - [1 - 8(-6)]$ Do the exponentiation.
    $= -5 + 3(4) - [1 + 48]$ Do the multiplication and
                                              subtraction inside the
                                              brackets.
    $= -5 + 3(4) - 49$
    $= -5 + 12 - 49$ Multiply.
    $= -42$ Add and subtract from left to
                                   right.

49. $9(x - 2) - 2[3(2x - 1) - 3(-2x + 3)]$
    $= 9(x - 2) - 2[6x - 3 + 6x - 9]$ Do the work inside the
                                          brackets first. Use the
                                          distributive law.
    $= 9(x - 2) - 2(12x - 12)$ Combine like terms.
    $= 9x - 18 - 24x + 24$ Use the distributive law.
    $= -15x + 6$ Combine like terms.

Evaluate the following if $x = -5$ and $y = 6$:

53. $9(x - 2y) - [3(2x - y) - 6(-3x + 2y)]$
    $= 9(x - 2y) - [6x - 3y + 18x - 12y]$ Use the distributive law
                                                inside the brackets.
    $= 9(x - 2y) - [24x - 15y]$ Combine like terms
                                         inside brackets.
    $= 9x - 18y - 24x + 15y$ Distributive law.
    $= -15x - 3y$ Combine like terms.
    $= -15(-5) - 3(6)$ Replace x with -5 and
                                    y with y.
    $= 75 - 18$
    $= 57$

57. $(2x - y)(x + y)(-3x - 4y)$
    $= (2 \cdot -5 - 6)(-5 + 6)(-3 \cdot -5 - 4 \cdot 6)$ Substitute -5 for x and
                                                       6 for y.
    $= (-10 - 6)(1)(15 - 24)$ Perform the indicated
    $= (-16)(1)(-9)$ operations.
    $= 144$

D

61. Use the formula $F = \frac{9}{5}C + 32$ to convert 50°C to degrees Fahrenheit.

$F = \frac{9}{5}C + 32$	Formula.
$F = \frac{9}{5} \cdot 50 + 32$	Substitute 50 for C.
$F = 90 + 32$	Perform the indicated operations.
$F = 122$	

So 150°C is equal to 122°F.

65. Find the total price of a TV if the down payment is $50 and there are 24 monthly payments of $29.18. (Use the formula T = D + pm, where T is the total price, D is the downpayment, p is the payment per month, and m is the number of months.)

T = D + pm	Formula.
T = 50 + 24(29.18)	Substitute 50 for D,
T = 50 + 700.32	29.18 for p, and 24 for m.
T = 750.32	

The total price of the TV is $750.32.

69. How far does an object fall after three seconds? (Use the formula $s = 16t^2$, where s represents the distance in feet and t represents time.)

$s = 16t^2$	Formula.
$s = 16 \cdot 3^2$	Replace t with 3.
$s = 16 \cdot 9$	
$s = 144$	

The object falls 144 ft.

EXERCISES 1.7   SOLVING EQUATIONS

A

Solve:

1.
$x + 2 = 5$
$x + 2 - 2 = 5 - 2$ Step 3. Subtract 2 from both sides.
$x = 3$

The solution set is {3}.

5. $-2x = -4$
    $\dfrac{-2x}{-2} = \dfrac{-4}{-2}$ Step 4. Divide both sides by -2.

    The solution set is $\{2\}$.

9. $-x + 1 = 3$
    $-x + 1 - 1 = 3 - 1$ Step 3. Subtract 1 from both sides of the equation.
    $-x = 2$ Step 4. Divide both sides by -1.
    $\dfrac{-x}{-1} = \dfrac{2}{-1}$
    $x = -2$

    The solution set is $\{-2\}$.

13. $x = x + 1$
    $x - x = x - x + 1$ Step 2. Subtract x from both sides of the equation. The equation is a contradiction, so there is no solution.
    $0 = 1$

    The solution set is empty, $\emptyset$.

B

17. $2x + 3 = 3 + 2x$
    $2x + 3 = 2x + 3$ Rewrite the right side of the equation, using the Commutative Law of Addition. The equation is an identity and so has all real numbers as its solution.

    The solution set is the set R, all real numbers.

21. $2x + 7 = 5x + 4 - 3x$
    $2x + 7 = 2x + 4$ Simplify.
    $2x - 2x + 7 = 2x - 2x + 4$ Step 2. Subtract 2 from each side of the equation.
    $7 = 4$ The equation is a contradiction, so there is no solution.

    The solution is the empty set, $\emptyset$.

25.
$$4x - 43 = 9x - 40$$
$$4x - 9x - 43 = 9x - 9x - 40 \qquad \text{Step 2. Subtract 9x from each side. Simplify.}$$
$$-5x - 43 = -40$$
$$-5x - 43 + 43 = -40 + 43 \qquad \text{Step 3. Add 43 to both sides. Simplify.}$$
$$-5x = 3$$
$$\frac{-5x}{-5} = \frac{3}{-5} \qquad \text{Step 4. Divide both sides by -5.}$$
$$x = -\frac{3}{5}$$

The solution set is $\left\{-\frac{3}{5}\right\}$.

29.
$$5x + 6 - (10 - 2x) = (7x + 6) - (6x - 2)$$
$$5x + 6 - 10 + 2x = 7x + 6 - 6x + 2 \qquad \text{Simplify on both sides by using the distributive law and collecting like terms.}$$
$$7x - 4 = x + 8$$
$$7x - 4 - x = x - x + 8 \qquad \text{Subtract x from both sides.}$$
$$6x - 4 = 8$$
$$6x - 4 + 4 = 8 + 4 \qquad \text{Add 4 to both sides.}$$
$$6x = 12$$
$$\frac{6x}{6} = \frac{12}{6} \qquad \text{Divide both sides by 6.}$$
$$x = 2$$

The solution set is $\{2\}$.

C

33.
$$2x - [3x - (8x + 2) + 3] + 6 = 0$$
$$2x - [3x - 8x - 2 + 3] + 6 = 0 \qquad \text{Simplify by performing}$$
$$2x - [-5x + 1] + 6 = 0 \qquad \text{indicated multiplications}$$
$$2x + 5x - 1 + 6 = 0 \qquad \text{and combining like terms.}$$
$$7x + 5 = 0$$
$$7x + 5 - 5 = 0 - 5 \qquad \text{Step 3. Subtract 5 from both sides.}$$
$$7x = -5$$
$$\frac{7x}{7} = \frac{-5}{7} \qquad \text{Step 4. Divide both side by 7.}$$
$$x = -\frac{5}{7}$$

The solution set is $\left\{-\frac{5}{7}\right\}$.

37.
$$x - \{3x - [2x - (8 - x) + 2] + 3\} - 6 = 0$$
$$x - \{3x - [2x - 8 + x + 2] + 3\} - 6 = 0 \qquad \text{Simplify by performing}$$
$$x - \{3x - [3x - 6] + 3\} - 6 = 0 \qquad \text{indicated multiplica-}$$
$$x - \{3x - 3x + 6 + 3\} - 6 = 0 \qquad \text{tions, and combining}$$
$$x - \{9\} - 6 = 0 \qquad \text{like terms.}$$
$$x - 9 - 6 = 0$$
$$x - 15 = 0$$
$$x - 15 + 15 = 0 + 15 \qquad \text{Step 3. Add 15 to each side.}$$
$$x = 15$$

The solution set is $\{15\}$.

41. $6x - 8(2x + 4) = 8x - 12(6 - x)$
     $6x - 16x - 32 = 8x - 72 + 12x$ Simplify by performing
     $-10x - 32 = 20x - 72$ indicated multiplications.
     $-10x - 20x - 32 = 20x - 20x - 72$ Then combine like terms.
     $-30x - 32 = -72$ Step 2. Subtract 20x from
                                              both sides.
     $-30x - 32 + 32 = -72 + 32$ Step 3. Add 32 to each side.
     $-30x = -40$
     $\dfrac{-30x}{-30} = \dfrac{-40}{-30}$ Step 4. Divide each side
                                              by -30.
     $x = \dfrac{4}{3}$

The solution set is $\left\{\dfrac{4}{3}\right\}$.

45. $-[3(x - 2) + 6] - [-2(x - 1) + 5] = -2[x + 4) - 1]$
     $-[3x - 6 + 6] - [-2x + 2 + 5] = -2[x + 4 - 1]$ Distributive law.
     $-[3x] - [-2x + 7] = -2[x + 3]$
     $-3x + 2x - 7 = -2x - 6$ Collect like
     $-x - 7 = -2x - 6$ terms.
     $-x + 2x - 7 = -2x + 2x - 6$ Add 2x to both
     $x - 7 = -6$ sides.
     $x - 7 + 7 = -6 + 7$ Add 7 to both
     $x = 1$ sides.

The solution set is {1}.

D

49. The cost of advertising for a minute on television is $250 more than 3 times the cost for minute on the radio. If an advertiser spends $1000 for one minute of television time and two minutes of radio time, what is the cost per minute of each?

Simpler word form:

$2 \begin{bmatrix} \text{Cost of one} \\ \text{minute of} \\ \text{radio} \\ \text{advertising} \end{bmatrix} + \begin{bmatrix} \text{Cost of one} \\ \text{minute of} \\ \text{television} \\ \text{advertising} \end{bmatrix} = 100$

Select variable:
Let c be the cost of one minute of radio advertising. Then $3c + 250$ is the cost of one minute of television advertising.

Translate to algebra:
$2c + (3c + 250) = 1000$

Solve:
$$2c + 3c + 250 = 1000$$
$$5c + 250 = 1000$$
$$5c = 750$$
$$c = 150$$

If $c = 150$, then $3 \cdot c + 250 = 3(150) + 250$
$$= 450 + 250$$
$$= 700$$

Each minute of radio time costs $150 and each minute of television time costs $700.

53. The cost of a television set is $50 more than twice the cost of a microwave oven. If the total cost for both is $650, what is the cost of the microwave?

    Simpler word form:
    $$\left[\begin{array}{c}\text{Cost of}\\ \text{television}\\ \text{set}\end{array}\right] + 2\left[\begin{array}{c}\text{Cost of}\\ \text{microwave}\\ \text{oven}\end{array}\right] + 50 = 650$$

    Select variable:
    Let x represent the cost of the microwave oven; then 2x + 50 represents the cost of the television set.

    Translate to algebra:
    $$(x) + (2x + 50) = 650$$

    Solve:
    $$3x + 50 = 650$$
    $$3x = 600 \quad \text{Subtract 50 from both sides.}$$
    $$x = 200 \quad \text{Divide both sides by 3.}$$

    The microwave oven costs $200.

57. One number is two more than three times another. If the sum of the two numbers is 22, find the two numbers.

    Simpler word form:
    (a number) + 3(another number) + 2 = 22

    Select a variable:
    Let x represent one number; then 3x + 2 represents the other number.

    Translate to algebra:
    $$(x) + (3x + 2) = 22$$

<u>Solve:</u>
$4x + 2 = 22$
$4x = 20$ Subtract 2 from both sides.
$x = 5$ Divide both sides by 4.

And $3x + 2 = 3(5) + 2$ Replace x with 5 to find the
$ = 15 + 2$ other number.
$ = 17$

One number is 5, and the other one 17.

EXERCISES 1.8 LITERAL EQUATIONS AND FORMULAS

A

Solve for the indicated variable:

1. $d = rt$; t

   $\dfrac{d}{r} = \dfrac{rt}{r}$ Divide both sides by r to isolate t.

   $\dfrac{d}{r} = t$

   $t = \dfrac{d}{r}$ Rewrite using the symmetric property of equality.

5. $V = \pi r^2 h$; h

   $\pi r^2 h = V$ Symmetric property of equality.

   $\dfrac{\pi r^2 h}{\pi r^2} = \dfrac{V}{\pi r^2}$

   $h = \dfrac{V}{\pi r^2}$

9. $F = \dfrac{GMm}{r^2}$; m

   $\dfrac{GMm}{r^2} = F$ Symmetric property of equality.

   $r^2 \left[\dfrac{GMm}{r^2}\right] = r^2(F)$ Multiply both sides by r^2.

   $GMm = Fr^2$

   $\dfrac{GMm}{GM} = \dfrac{Fr^2}{GM}$ Divide both sides by GM.

   $m = \dfrac{Fr^2}{GM}$

13.
$$y = mx - 3;\ x$$
$$mx - 3 = y$$ Symmetric property of equality.
$$mx - 3 + 3 = y + 3$$ To isolate the term containing the variable x, add 3 to each side.
$$mx = y + 3$$
$$\frac{mx}{m} = \frac{y + 3}{m}$$ Divide both sides by m.
$$x = \frac{y + 3}{m}$$

B

17.
$$ax - 5y = 2;\ x$$
$$ax - 5y + 5y = 5y + 2$$ To isolate the term containing x, add 5y to both sides.
$$ax = 5y + 2$$
$$\frac{ax}{a} = \frac{5y + 2}{a}$$ Divide both sides by a.
$$x = \frac{5y + 2}{a}$$

21.
$$PV = nRT;\ n$$
$$nRT = PV$$ Symmetric property of equality.
$$\frac{nRT}{RT} = \frac{PV}{RT}$$ Divide both sides by RT.
$$n = \frac{PV}{RT}$$

25.
$$A = \frac{B - C}{n};\ B$$
$$\frac{B - C}{n} = A$$ Symmetric property of equality.
$$n\left[\frac{B - C}{n}\right] = n(A)$$ Multiply both sides to eliminate the fraction.
$$B - C = An$$
$$B - C + C = An + C$$ Add C to both sides.
$$B = An + C$$

29.
$$L = f + (n - 1)d;\ d$$
$$f + (n - 1)d = L$$ Symmetric property of equality.
$$f - f + (n - 1)d = L - f$$ Subtract f from each side.
$$(n - 1)d = L - f$$
$$\frac{(n - 1)d}{(n - 1)} = \frac{(L - f)}{(n - 1)}$$ Divide both sides by n - 1.
$$d = \frac{(L - f)}{(n - 1)}$$

33. $\quad P = 2L + 2W; \ L$

$$2L + 2W = P$$

Symmetric property of equality.

$$2L + 2W - 2W = P - 2W$$
$$2L = P - 2W$$

Subtract 2W from both sides isolate the term with L.

$$\frac{2L}{2} = \frac{P - 2W}{2}$$

Divide both sides by 2.

$$L = \frac{P - 2W}{2}$$

37. $\quad s = \frac{1}{2}gt^2 + vt + x; \ g$

$$\frac{1}{2}gt^2 + vt + x = s$$

Symmetric property of equality.

$$\frac{1}{2}gt^2 = s - vt - x$$

Subtract vt and x from both sides, isolating the term with g.

$$2\left[\frac{1}{2}gt^2\right] = 2(s - vt - x)$$
$$gt^2 = 2s - 2vt - 2x$$

Multiply both sides by 2 to clear the fraction on the left.

$$\frac{gt^2}{t^2} = \frac{2s - 2vt - 2x}{t^2}$$

Divide both sides by t^2.

$$g = \frac{2s - 2vt - 2x}{t^2}$$

41. $\quad y = \frac{4}{5}x - 5; \ x$

$$\frac{4}{5}x - 5 = y$$

Symmetric property of equality.

$$\frac{4}{5}x - 5 + 5 = y + 5$$

Add 5 to both sides.

$$\frac{4}{5}x = y + 5$$

$$\frac{5}{4}\left[\frac{4}{5}x\right] = \frac{5}{4}\left[\frac{y + 5}{1}\right]$$

Multiply both sides by the reciprocal of the coefficient of x.

$$x = \frac{5(y + 5)}{4}$$

$$x = \frac{(5y + 25)}{4}$$

Simplify.

45. $\quad 2A = h(b + B); \ B$

$$h(b + B) = 2A$$

Symmetric property of equality.

$$bh + Bh = 2A$$
$$bh - bh + Bh = 2A - bh$$
$$Bh = 2A - bh$$

Eliminate the parentheses by doing the indicated multiplication. Subract bh from each side.

$$\frac{Bh}{h} = \frac{2A - bh}{h}$$

Divide both sides by h.

$$B = \frac{(2A - bh)}{h}$$

Alternate solution:
$$2A = h(b + B)$$
$$h(b + B) = 2A$$ Symmetric property of equality.

$$\frac{h(b + B)}{h} = \frac{2A}{h}$$ Divide both sides by h.

$$b + B = \frac{2A}{h}$$

$$b - b + B = \frac{2A}{h} - b$$ Subtract b from both sides.

$$B = \frac{2A}{h} - b$$

So $B = \frac{2A - bh}{h}$ or $\frac{2A}{h} - b$ Either form is acceptable.

EXERCISES 1.9   SOLVING INEQUALITIES

A

Solve and graph the solution set:

1. $x < -5$

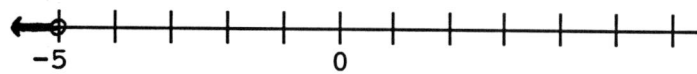

The circle on -5 is open since -5 is not a member of the solution set.

5. $-y \leq 2$

$$\frac{-y}{-1} \geq \frac{2}{-1}$$ Divide both sides by -1. The inequality reverses.

$$y \geq -2$$

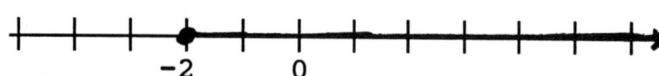

The graph contains all points to the right of -2. -2 is included in the solution set, and so it is shaded.

9. $\quad -x + 4 > 0$
$$-x + 4 - 4 > 0 - 4$$ Subtract 4 from each side.
$$-x > -4$$
$$\frac{-x}{-1} < \frac{-4}{-1}$$ Divide both sides by -1. The inequality reverses.
$$x < 4$$

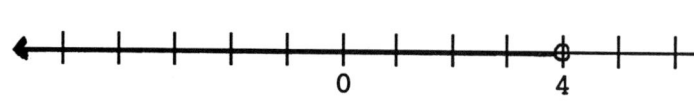

The graph contains all points to the left of 4. 4 is not included in the solution set and so it is unshaded.

Solve and write the solution set in interval notation:

13. $2x < 8$

    $\dfrac{2x}{2} < \dfrac{8}{2}$ Divide both sides by 2.

    $x < 4$

    The solution set is $(-\infty, 4)$.   The end point, 4, is not included in the solution, so a parenthesis is used instead of a bracket.

B

Solve and write in set builder notation:

17. $2x + 3 < 13$

    $2x + 3 - 3 < 13 - 3$ Subtract 3 from each side.

            $2x < 10$

            $\dfrac{2x}{2} < \dfrac{10}{2}$ Divide both sides by 2.

             $x < 5$

    The solution set is $\{x \mid x < 5 \text{ and } x \in \mathbb{R}\}$.

Solve and graph the solution set:

21. $5a - 9 > 21$

    $5a - 9 + 9 > 21 + 9$ Add 9 to both sides.

            $5a > 30$

            $\dfrac{5a}{5} > \dfrac{30}{5}$ Divide each side by 5.

             $a > 6$

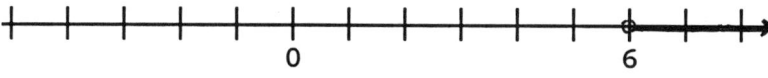

25. $-x + 7 > x + 5$

    $-x - x + 7 > x - x + 5$ Subtract x from each side.

            $-2x + 7 > 5$

    $-2x + 7 - 7 > 5 - 7$ Subtract 7 from each side.

              $-2x > -2$

              $\dfrac{-2x}{-2} > \dfrac{-2}{-2}$ Divide each side by -2. The inequality reverses.

               $x < 1$

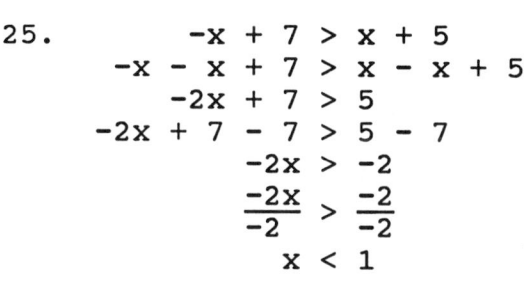

Solve and write the solution set in set builder notation:

29. $-5(2x + 3) \geq 4(x - 2)$
     $-10x - 15 \geq 4x - 8$ Distributive property.
     $-14x - 15 \geq -8$ Subtract 4x from both sides.
          $-14x \geq 7$ Add 15 to both sides.
          $\dfrac{-14x}{-14} \leq \dfrac{7}{-14}$ Divide each side by -14, reversing the inequality.
             $x \leq -\dfrac{1}{2}$

The solution set is $\left\{ x \mid x \leq -\dfrac{1}{2} \right\}$

C

Solve and graph the solution set:

33. $5x - (3x + 2) \geq x - 9$
      $5x - 3x - 2 \geq x - 9$ Do the indicated multiplication.
          $2x - 2 \geq x - 9$ Combine the like terms on
     $2x - x - 2 \geq x - x - 9$ the left.
            $x - 2 \geq -9$
         $x - 2 + 2 \geq -9 + 2$ Add 2 to each side.
              $x \geq -7$

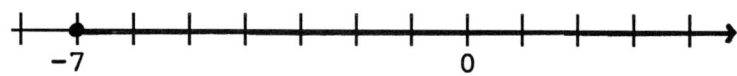

37. $4x - [5 - (2x - 3)] - 6(x + 5) \geq 10$
      $4x - [5 - 2x + 3] - 6(x + 5) \geq 10$ Distributive property inside brackets.
        $4x - [-2x + 8] - 6(x + 5) \geq 10$ Combine like terms.
          $4x + 2x - 8 - 6x - 30 \geq 10$ Do the indicated multiplications.
                               $-38 \geq 10$ Combine like terms.

Since -38 is not greater than or equal to 10, this equation is a contradiction, and the solution set is the empty set of { } or ∅.

Solve and write the solution set in interval notation:

41. $-12 \leq 5x - 3 < 22$
   $-12 + 3 \leq 5x - 3 + 3 < 22 + 3$ Add 3 to each member.
   $-9 \leq 5x < 25$
   $\frac{-9}{5} \leq \frac{5x}{5} > \frac{25}{5}$ Divide each member by 5.
   $-\frac{9}{5} \leq x < 5$

   The solution set is $[-\frac{9}{5}, 5)$.   The bracket indicates that $-\frac{9}{5}$ is included in the solution set, while the parenthesis indicates that 5 is not included in the solution set.

45. $25 \leq 7x + 3 \leq 66$
   $25 - 3 \leq 7x + 3 - 3 \leq 66 - 3$ Subtract 3 from each member.
   $22 \leq 7x \leq 63$
   $\frac{22}{7} \leq \frac{7x}{7} \leq \frac{63}{7}$ Divide each member by 7.
   $\frac{22}{7} \leq x \leq 9$

   The solution set is $[\frac{22}{7}, 9]$.   The brackets indicate that both endpoints are included in the solution set.

D

49. The charges for a bus tickets are $20 plus 12¢ per mile. Could a trip of 800 miles be taken if the person had at the most $120 to spend?

   Simpler word form:
   $20 + $.12 for each mile traveled is less than or equal to $120.

   Select variable:
   Let x represent the number of miles to be traveled. Then 0.12x represents the cost of x miles of travel.

<u>Translate to algebra:</u>
$20 + 0.12x \leq 120$

<u>Solve:</u>
$$20 - 20 + 0.12x \leq 120 - 20$$
$$0.12x \leq 100$$
$$\frac{0.12x}{0.12} \leq \frac{100}{0.12}$$
$$x \leq 833\tfrac{1}{3}$$

<u>Answer:</u>

Since the trip to be taken is less than $833\tfrac{1}{3}$ miles long, it will be possible to make the trip and spend at most $120.

53. Jack can spend, at most, $55 on paint. If a gallon can of paint costs $7.50 per can, what is the maximum number of cans of paint Jack can purchase?

<u>Simpler word form:</u>
$$\begin{bmatrix} \text{cost} \\ \text{per} \\ \text{can} \end{bmatrix} \cdot \begin{bmatrix} \text{number} \\ \text{of} \\ \text{cans} \end{bmatrix} \leq 55$$

<u>Select a variable:</u>
Let x represent the number of cans to be purchased. Note that x is an integer.

<u>Translate to algebra:</u>
$7.50x \leq 55$

<u>Solve:</u>
$$\frac{7.50x}{7.50} \leq \frac{55}{7.50}$$
$$x \leq 7\tfrac{1}{3}$$

Divide both sides by 7.50 to isolate the variable.

The greatest integer less than $7\tfrac{1}{3}$ is 7.

Jack can buy 7 cans of paint.

57. It costs $3.50 for the first pound and $1.75 per pound after the first pound for a messenger to deliver a package. Will Charlie be able to get his 5-lb package delivered if he has only $10.00?

$$3.50 + 4(1.75) \leq 10$$
$$3.50 + 7.00 \leq 10$$

Find the sum of the cost of the first pound (3.50) and the cost of each subsequent pound.

$$10.50 \leq 10$$

Since the cost $10.50 is greater than $10.00, this statement is false.

No, Charlie will be unable to get the package delivered.

EXERCISES 1.10   WORD PROBLEMS

1. The length of a rectangle is 4 times its width. The perimeter is 78 feet. Find the length and width.

   <u>Select a variable:</u>
   Let w represent width. Then the length is 4 times the width, or 4 times w, or 4w.

   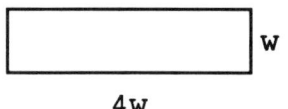
   4w

   Make a drawing.
   Label the length and width using the variable selected.

   <u>Formula:</u>
   $P = 2\ell + 2w$

   Formula for the perimeter of a rectangle.

   $P = 78, \ell = 4w$
   $78 = 2(4w) + 2w$

   Substitute into the formula.

   <u>Solve:</u>
   $78 = 8w + 2w$
   $78 = 10w$
   $7.8 = w$

   So the length is 4(7.8) or 31.2 ft.

   <u>Answer:</u>
   The width is 7.8 ft, and the length is 31.2 ft.

5. Two cars leave Austin traveling in opposite directions. If their average speeds are 42 mph and 54 mph, how long will it be until they are 312 miles apart?

   The problem states the speeds of the two cars and the distance between them.

   <u>Simpler word form:</u>
   $$\begin{bmatrix} \text{Distance} \\ \text{traveled} \\ \text{by Car 1} \end{bmatrix} + \begin{bmatrix} \text{Distance} \\ \text{traveled} \\ \text{by Car 2} \end{bmatrix} = \text{Total Distance}$$

   <u>Select variable:</u>
   Let T represent the number of hours until the cars are 312 miles apart.

	Rate (Speed)	Time	Distance
Car 1	42	t	42t
Car 2	54	t	54t

   Make a table.

   <u>Translate to algebra:</u>
   42t + 54t = 312

   <u>Solve:</u>
   42t + 54t = 312
   96t = 312
   t = 3.25 = $3\frac{25}{100}$ = $3\frac{1}{4}$

   <u>Answer:</u>
   In $3\frac{1}{4}$ hrs (or 3 hrs and 15 min) the cars will be 312 miles apart.

9. Myrna's baby girl weighed 5.75 pounds at birth. At age 7 months she weighed 16.25 pounds. What was the percent of increase in the 7 months (to the nearest percent)?
   The problem gives two weights.
   The question asks for a comparison (in percent) between the increase and the first weight.

   <u>Simpler word form:</u>
   What percent of 5.75 is 10.5?   Note that the increase is 10.5. (16.25 - 5.75)

   <u>Select a variable:</u>
   Let x represent the percent.

   <u>Translate to algebra:</u>
   x(5.75) = 10.5

Solve:
$$\frac{x(5.75)}{5.75} = \frac{10.5}{5.75}$$
$$x \approx 1.83 \approx 183\%$$

Answer:
The baby's weight increased approximately 183%.

13. The first side of a triangle is 15 inches. The third side is 3 less than 2 times the second side. If the perimeter is 51 inches, find the lengths of the second and third sides.

    Simpler word form:
    $$\begin{bmatrix}\text{first}\\\text{side}\end{bmatrix} + \begin{bmatrix}\text{second}\\\text{side}\end{bmatrix} + \begin{bmatrix}\text{third}\\\text{side}\end{bmatrix} = 51 \quad \text{Formula for the perimeter of a triangle.}$$

    Select a variable:
    Let s represent the second side. Then 2s - 3 represents the third side.

    Translate to algebra:
    (15) + (s) + (2s - 3) = 51

    Solve:
15 + 3s - 3 = 51	Combine like terms.
3s + 12 = 51	
3s = 39	Subtract 12 from both sides.
x = 13	Divide both sides by 3.

    So 2s - 3 = 2(13 - 3
              = 26 - 3
              = 23

    Answer:
    The lengths of the second and third sides are 13 in. and 23 in., respectively.

30

CHAPTER 2

POLYNOMIALS

EXERCISES 2.1 POSITIVE INTEGER EXPONENTS AND MULTIPLICATION OF MONOMIALS

A

Multiply:

1. $r^7 \cdot r^4 \cdot r = r^{7+4+1}$
   $= r^{12}$

   Law I. Add the exponents. Remember that $r = r^1$.

5. $(a^4)^4 = a^{4 \cdot 4}$
   $= a^{16}$

   Law II. Multiply the exponents.

9. $(-3x^2y^3)^2 = (-3)^2(x^2)^2(y^3)^2$

   $= 9x^4y^6$

   Law III. Raise each factor in parentheses, independently, to the second power.
   Law II. Multiply the exponents when raising a power to a power.

13. $(x^3y)(xy^3)(x^2y^2)$
    $= [x^3 \cdot x \cdot x^2][y \cdot y^3 \cdot y^2]$
    $= x^6y^6$

    Group the factors that can be simplified together.
    Law I. Add the exponents and keep the common base.

B

17. $(2x^{11})(-3x^{12})(5x^{15})$
    $= [2 \cdot -3 \cdot 5][x^{11}x^{12}x^{15}]$
    $= -30x^{38}$

    Group the factors that can be simplified together.
    Law I. Add the exponents.

21. $(-14p^2q^3)^2 = (-14)^2(p^2)^2(q^3)^2$

    $= 196p^4q^6$

    Law III. Raise each factor in parentheses, independently, to the second power.
    Law II. Multiply the exponents when raising a power to a power.

25. $(t^2)^3(t^3)^2$
    $= t^6 \cdot t^6$

    $= t^{12}$

    Law II. Multiply the exponents when raising a power to a power.
    Law I. Add the exponents.

29. $(-2x)^2(-2x)^3$
    $= (-2)^2(x)^2(-2)^3(x)^3$

    $= (4)(x^2)(-8)(x^3)$
    $= (4 \cdot -8)(x^2 \cdot x^3) = -32x^5$

Law III. Raise each factor in parentheses to the required power.

Regroup and simplify.

C

33. $(-2xy)^3(4x^2y^3)$
    $= [(-2)^3x^3y^3](4x^2y^3)$
    $= (-8x^3y^3)(4x^2y^3)$

    $= [(-8)(4)][x^3 \cdot x^2][y^3 \cdot y^3]$

    $= -32x^5y^6$

Law III. In first term, raise each factor of $-2 \cdot x \cdot y$ to the third power.
Group the constants and group the variables.
Law I. Add the exponents on each base.

37. $(2t)^2(-3t^3)^3(-t^2)^4$
    $= (2)^2(t)^2(-3)^3(t^3)^3(-1)^4(t^2)^4$

    $= [4 \cdot -27][t^2 t^9 \cdot t^8]$

    $= -108t^{19}$

Raise each factor in each set of parentheses to the required power.
Group the factors that can be simplified together and use Law II.

41. $(x + y)^2(x + y)^4$
    $= (x + y)^{2+4} = (x + y)^6$

The common base is $(x + y)$.
Law I. To multiply two exponential expressions with a common base, add the exponents.

45. $(2t^{4n})(t)^{2n-2}$
    $= (2)[t^{4n} \cdot t^{2n-2}]$

    $= 2t^{6n-2}$

Group the constants and group the variables.

Law I. Add the exponents of the base t: $4n + (2n - 2)$.

D

49. If a special cylinder has a height that is the fifth power of the radius ($h = r^5$), express the volume of the cylinder in terms of the radius. (Use the formula in the application.)

    Formula:
    $V = \pi r^2 h$

    Formula for the volume of a cylinder.

    Substitute:
    $V = \pi r^2(r^5)$
    $V = \pi r^7$

    Replace h by r^5 and multiply.

53. The area of a circle is $A = \pi r^2$, where r is the radius of the circle. The radius is half the diameter, $r = \frac{1}{2}d$. Find a formula for the area in terms of the diameter.

$A = \pi r^2$ Formula.

$A = \pi \left[\frac{1}{2}d\right]^2$ Substitute $\frac{1}{2}d$ for r.

$A = \pi \left[\frac{1}{4}d^2\right]$ Law III.

$A = \frac{\pi d^2}{4}$ Simplify.

57. The volume of a cone is given by the formula $V = \frac{1}{3}\pi r^2 h$, where r is the radius and h is the height. If the radius is three times the cube of the height, find an expression for the volume of the cone in terms of the height.

$V = \frac{1}{3}\pi r^2 h$

$V = \frac{1}{3}\pi (3h^3)^2 (h)$ The radius is three times the cube of the height, or $3h^3$ so replace r with $3h^3$.

$V = \frac{1}{3}\pi (9h^6)(h)$ Law II: $(3h^3)^2 = 3^2 h^6$.

$V = \frac{1}{3}\pi (9h^7)$ Law I: $h^6 \cdot h = h^7$

$V = \frac{9\pi h^7}{3} = 3\pi h^7$

E   Maintain Your Skills

Simplify (add, subtract, or combine like terms):

61. $(-18.4) + (-21.7) - (13.84) - (-21.52)$
 $= (-18.4) + (-21.7) + (-13.84) + (21.52)$ Subtract by adding the opposite.
 $= (-40.1) + (-13.84) + (21.52)$ Perform the indicated operations from left to right.
 $= (-54.94) + (21.52)$
 $= -32.42$

65. $76ab - 19bc - 71ab - 67bc$
 $= (76ab - 71ab) + (-19bc - 67bc)$ Group like terms.
 $= 5ab - 86bc$ Combine like terms.

EXERCISES 2.2 INTEGER EXPONENTS AND DIVISION OF MONOMIALS

A

1. $8^{-2} = \frac{1}{8^2} = \frac{1}{64}$

The negative sign denotes the reciprocal.

5. $a^{-1}b^2 = \frac{1}{a}\cdot b^2 = \frac{b^2}{a}$

Write each factor. Use the reciprocal of each factor that has a negative exponent.

9. $\frac{a^{-3}b}{a^{-7}b^{-2}} = a^{-3-(-7)}b^{1-(-2)}$

$= a^4b^3$

Divide using Law IV.

13. $(a^{-1}b^2)^{-2} = a^2b^{-4}$

$= a^2\cdot\frac{1}{b^4} = \frac{a^2}{b^4}$

Raise each factor in parentheses to the -2 power. Write the reciprocal of each factor that has a negative exponent.

B

17. $(2x^{-3}y^2)^2(-6x^2y^{-4})$

$= -12x^{-3+2}y^{2+(-4)} = -12x^{-1}y^{-2}$

$= -12\cdot\frac{1}{x}\cdot\frac{1}{y^2}$

$= -\frac{12}{xy^2}$

Multiply powers with a common base. Law I.
Write the reciprocal of each factor that has a negative exponent.

21. $(2x^2y^{-3})(-3x^{-2}y)^3$

$= (2x^2y^{-3})(-3)^{-3}(x^{-2})^{-3}(y)^{-3}$

$= 2x^2y^{-3}(-3)^{-3}x^6y^{-3}$

$= 2(-3)^{-3}x^{2+6}y^{-3+(-3)}$

$= 2(-3)^{-3}x^8y^{-6}$

$= 2\cdot\frac{1}{(-3)^3}\cdot x^8\cdot\frac{1}{y^6}$

$= 2\cdot\frac{1}{-27}\cdot x^8\cdot\frac{1}{y^6}$

$= -\frac{2x^8}{27y^6}$

Raise each factor in the second set of parentheses to the -3 power.
Rewrite using Law II on $(x^2)^{-3}$.
Use Law I to simplify factors having a common base.

Write the reciprocals of factors having a negative exponent.

Simplify.

25. $\left[\dfrac{m^{-1}}{n}\right]^{-1} = \dfrac{(m^{-1})^{-1}}{(n)^{-1}}$ Raise both numerator and denominator to the power -1, independently (Law V).

$= \dfrac{m^{(-1)(-1)}}{n^{-1}} = \dfrac{m^1}{n^{-1}}$

$= m \cdot \dfrac{n}{1} = mn$

29. $\left[\dfrac{a^5}{b^2}\right]^{-4} = \dfrac{(a^5)^{-4}}{(b^2)^{-4}}$ Raise both numerator and denominator to the power -4, independently.

$= \dfrac{a^{5(-4)}}{b^{2(-4)}} = \dfrac{a^{-20}}{b^{-8}}$

$= \dfrac{1}{a^{20}} \cdot \dfrac{b^8}{1} = \dfrac{b^8}{a^{20}}$

33. $\left[\dfrac{x^2}{3y^{-3}}\right]^2 = \dfrac{(x^2)^2}{(3y^{-3})^2}$ Raise both numerator and denominator to the power 2, independently.

$= \dfrac{x^4}{3^2 y^{-6}} = \dfrac{x^4}{9y^{-6}}$

$= \dfrac{x^4 y^6}{9}$ The reciprocal of $\dfrac{1}{y^{-6}}$ is y^6.

37. $\left[\dfrac{5xy^6}{3z^4}\right]^{-2} = \left[\dfrac{3z^4}{5xy^6}\right]^2$ Since the exponent is negative, write the reciprocal of the quantity in parentheses.

$= \dfrac{3^2 z^8}{5^2 x^2 y^{12}}$

$= \dfrac{9z^8}{25 x^2 y^{12}}$

41. $\dfrac{(6x^2)^{-2}(3x^3)^2}{(2x^3)^{-4}}$

$= \dfrac{(6^{-2}x^{-4})(3^2x^6)}{2^{-4}x^{-12}}$ Law III.

$= \dfrac{3^2x^6 2^4 x^{12}}{6^2 x^4}$ Write each factor with a negative exponent as a reciprocal.

$= \dfrac{9 \cdot 16 \cdot x^6 \cdot x^{12}}{36 \cdot x^4}$

$= \dfrac{4x^{18}}{x^4} = 4x^{14}$

45. $\left[\dfrac{p^{-1}}{q^{-2}}\right]^{-1} \left[\dfrac{p^2}{q^{-1}}\right]^2 \left[\dfrac{p^{-1}q^{-1}}{p^{-2}}\right]^{-2}$

$= \left[\dfrac{q^{-2}}{p^{-1}}\right]^1 \left[\dfrac{p^2}{q^{-1}}\right]^2 \left[\dfrac{p^{-2}}{p^{-1}q^{-1}}\right]^2$ Rewrite each parenthetical expression that has a negative exponent as a reciprocal.

$= \dfrac{q^{-2}}{p^{-1}} \cdot \dfrac{p^4}{q^{-2}} \cdot \dfrac{p^{-4}}{p^{-2}q^{-2}}$ Law V.

$= \dfrac{p}{q^2} \cdot p^4 q^2 \cdot \dfrac{p^2 q^2}{p^4}$ Rewrite each factor that has a negative exponent as a reciprocal.

$= \dfrac{p^7 q^4}{p^4 q^2}$ Law I.

$= p^3 q^2$ Law IV.

49. $\dfrac{(x+3)^{-2}}{(x-4)^{-1}} \cdot \dfrac{(x-4)^2}{(x+3)^{-3}}$

$= \dfrac{(x-4)^1}{(x+3)^2} \cdot (x-4)^2 (x+3)^3$ Rewrite expressions with a negative exponent as reciprocals.

$= \dfrac{(x-4)^1 (x-4)^2 (x+3)^3}{(x+3)^2} = (x-4)^3 (x+3)$

36

53. A solution with pH of 5.3 has a hydrogen ion concentration of 4.9×10^{-6}. Write this concentration in decimal form.

4.9×10^{-6}
$= 4.9 \times \dfrac{1}{10^6}$ Write 10^{-6} as the reciprocal of 10^6.
$= 4.9 \times \dfrac{1}{1000000}$
$= 4.9 \times 0.000001$ Then write in decimal form.
$= 0.0000049$ This is the concentration in decimal form.

57. The wave length of an x-ray is 1×10^{-6} centimeters. Write this length in decimal form.

$1 \times 10^{-6} = 1 \times \dfrac{1}{10^6}$ Write 10^{-6} as the reciprocal of 10^6.
$= 1 \times \dfrac{1}{1000000}$
$= 1 \times 0.000001$ Then write in decimal form.
$= 0.000001$ This is the wave length in decimal form.

E   Maintain Your Skills

Solve:

61. $4(x - 5) - 6(5 - 3x) = 5(4x - 6)$
      $4x - 20 - 30 + 18x = 20x - 30$ Distributive law.
               $22x - 50 = 20x - 30$ Combine like terms.
                         $2x = 20$
                           $x = 10$

65. $4 - \{x - [x - (x - 4)]\}$
      $= 4 - \{x - [x - x + 4]\}$ Simplify the expression inside brackets.
      $= 4 - \{x - [4]\}$
      $= 4 - \{x - 4\}$ Simplify the expression inside braces.
      $= 4 - x + 4$
      $= -x + 8$ or $8 - x$

EXERCISES 2.3 SCIENTIFIC NOTATION

A

Write in scientific notation:

1. 50,000
   5.0 is between 1 and 10.   Step 1. Move the decimal point 4 places to the left.

   5.0 times 10000 or 10^4 = 50,000   Step 2. Multiplying by the fourth power of 10 gives the original value.

   50,000 = 5.0 × 10^4 or 5 × 10^4 Step 3.

5. 430,000
   4.3 is between 1 and 10.   Step 1. Move the decimal point 5 places to the left.

   4.3 times 100000 or 10^5 = 430,000   Step 2. Multiplying by the fifth power of 10 gives the original value.

   430,000 = 4.3 × 10^5 Step 3.

Change to place value notation:

9. 9.3×10^2 = 930

   Move the decimal point two places to the right. This is the equivalent of multiplying by 10 twice.

13. 2.32×10^{-3} = 0.00232

   Move the decimal point three places to the left. This is equivalent to multiplying by $\frac{1}{10}$ three times.

B

Write in scientific notation:

17. 377,000
    3.77 is between 1 and 10   Step 1.
    3.77 × 100000 or 10^5 = 377,000   Step 2.
    377,000 = 3.77 × 10^5 Step 3.

21. 611,000,000
    6.11 is between 1 and 10
    6.11 × 100000000 or 10^8 = 611,000,000   Move the decimal place 8 places to the left. Then multiply by the eighth power of 10.
    611,000,000 = 6.11 × 10^8

Change to place value notation:

25. $6.89 \times 10^4 = 68{,}900$ Move the decimal point 4 places to the right.

Perform the indicated operations using scientific notation and the laws of exponents. Write the result in scientific notation:

29. $(5.4 \times 10^{-3})(2 \times 10^{-4})$
     $= (5.4 \times 2)(10^{-3} \times 10^{-4})$ Use the commutative and distributive laws of multiplication to group the decimals and powers of ten.

   $= 10.8 \times 10^{-7}$ Multiply the decimal and whole number and add the exponents of 10.

   $= 1.08 \times 10^{-7} \times 10^1$ Note that 10.8 is not in scientific notation. Move the decimal point one place to the left and multiply by 10^1 to restore the original value.

   $= 1.08 \times 10^{-6}$ Law I of exponents.

33. $\dfrac{8.1 \times 10^{-5}}{3.0 \times 10^2} = \dfrac{8.1}{3.0} \times \dfrac{10^{-5}}{10^2}$

   $= 2.7 \times 10^{-7}$ Divide the decimals and subtract the exponents of 10.

C

Write in scientific notation:

37. $3784 = 3.784 \times 10^3$ Move the decimal 3 places to the left. Then multiplying by 10^3 restores the original value.

41. $0.000484 = 4.84 \times 10^{-4}$ Move the decimal 4 places to the right and then multiply by 10^{-4}.

Change to place value notation:

45. $2.36 \times 10^{-9} = 0.00000000236$ Move the decimal point 9 places to the left, which is equivalent to multiplying by $\frac{1}{10}$ nine times.

Perform the indicated operations using scientific notation and the laws of exponents. Write the results in both scientific notation and place value form:

49. $\dfrac{(7.2 \times 10^2)(3.6 \times 10^{-5})}{(4 \times 10^{-1})(5.4 \times 10^3)}$

$= \dfrac{25.92 \times 10^{-3}}{21.6 \times 10^2}$

In the numerator, multiply the decimals and add the exponents of 10. Do the same in the denominator. Now divide the decimals and subtract the exponents of 10. Move the decimal 5 places to the left, which is equivalent to multiplying by $\dfrac{1}{10}$ five times.

$= 1.2 \times 10^{-5}$

$= 0.000012$

D

Write in scientific notation:

53. In one year, light travels approximately 5,870,000,000,000 miles.

5,870,000,000,000 mi = 5.87×10^{12} mi   Move the decimal 12 places to the left. Then multiply by 10^{12}.

57. The length of a long x-ray is approximately 0.000001 centimeter.

0.000001 cm = 1.0×10^{-6} cm
= 1×10^{-6} cm

Move the decimal 6 places to the right, and then multiply by 10^{-6} to restore original value.

Write in place value notation:

61. The distance from Earth to the nearest star is approximately 2.55×10^{13} miles.

2.55×10^{13} mi = 25,500,000,000,000 mi

Move the decimal 13 places to the right, which is equivalent to multiplying by 10 thirteen times.

E   Maintain Your Skills

Add:

65. (13) + (-56) + (-102) + 43 + (-14)
   = (-43) + (-102) + 43 + (-14)   Perform indicated operations
   = (-145) + 43 + (-14)   from left to right.
   = (-102) + (-14)
   = -116

Combine terms:

69. 0.94x - 0.73y + 0.91 - 0.25x + 1.03y + 2.11 + 0.33x - 2.05y + 1.33
   = (0.94x - 0.25x + 0.33x) + (-0.73y + 1.03y - 2.05y)
       + (0.91 + 2.11 + 1.33)
   = 1.02x - 1.75y + 4.35

Solve:

73. The population of Hillsboro went from 18,750 to 30,000 in ten
     years. What was the percent of increase?

                                    Note that the actual increase
                                    in population was
                                    30,000 - 18,750 or 11,250.

   Simpler word form:
   11250 is what percent of 18750?

   Select a variable:
   Let x represent the percent of increase.

   Translate to algebra:
   11250 = 18750x

   Solve:
       x = 0.6 or 60%

   Answer:
   The percent of increase was 60%.

41

EXERCISES 2.4 POLYNOMIALS (PROPERTIES AND CLASSIFICATIONS)

A - D

Classify the following polynomials by number of terms and by degree:

Polynomial	Number of Terms	Degree
1. $4x + 2$	2 binomial	1 (linear)
5. 16	1 monomial	0 (zero)
9. $13x^3 - 4$	2 binomial	3 (cubic)
13. $\frac{3}{4}x - \frac{5}{8}y + 9z$	3 trinomial	1 (linear)

E Maintain Your Skills

True or False?

17. $-\frac{22}{3} < -\frac{36}{5}$

$-\frac{110}{15} < -\frac{108}{15}$

True.

Rewrite each fraction with 15 as the denominator. $-\frac{110}{15}$ is to the left of $-\frac{108}{15}$ on the number line.

21. The following is an example of what property of real numbers?

$-\frac{4}{5}\left(\frac{7}{7}\right) = -\frac{4}{5}$

Multiplication property of one.

$a \cdot 1 = a$

25. What is the reciprocal of -3.8?

$-3.8 = -3\frac{8}{10} = -\frac{38}{10}$

$-\frac{10}{38} = -\frac{5}{19}$

So $(-3.8)\left(-\frac{5}{19}\right) = 1$.

Rewrite -3.8 as a mixed number. Change the mixed number to an improper fraction.
Invert $-\frac{38}{10}$ and reduce.

Therefore, the reciprocal of -3.8 is $-\frac{5}{19}$.

EXERCISES 2.5   ADDITION AND SUBTRACTION OF POLYNOMIALS

A

Add or subtract as indicated:

1. $(4x + 3y) + (6x + 12y)$
    $(4x + 6x) + (3y + 12y)$ Group the like terms.
    $10x + 15y$ Combine like terms.

5. $(19x - 20) - (11x + 16)$
    $(19x - 20) + (-11x - 16)$ Use the definition of
                                             subtraction to change to
                                             addition.
    $(19x - 11x) + (-20 - 16)$ Group the like terms.
    $8x - 36$ Combine like terms.

Subtract:

9. $16x^2 + 5x + 20$
    $\underline{5x^2 + 2x + 6}$
    $11x^2 + 3x + 14$ Subtract by adding the
                                             opposite of the term being
                                             subtracted.

Add:

13. $(2.7p - 2.4q + 8.4) + (-3.8p + 3.1q - 2.7)$
    $(2.7p - 3.8p) + (-2.4q + 3.1q) + (8.4 - 2.7)$ Group like terms
    $-1.1p + 0.7q + 5.7$ and combine.

B

Add or subtract as indicated:

17. $(5x^3 + 4x^2 - 5x) + (3x^2 + 2x + 1) - (8x - 5)$
    $(5x^3 + 4x^2 - 5x) + (3x^2 + 2x + 1) + (-8x + 5)$ Use the definition
                                                         of subtraction to
                                                         change the
                                                         subtraction to
                                                         addition.
    $(5x^3) + (4x^2 + 3x^2) + (-5x + 2x - 8x) + (1 + 5)$ Group like terms.
    $5x^3 + 7x^2 - 11x + 6$ Combine like
                                                           terms.

Add:

21. $4a^2 - 3ab + b^2$, $12ab - 7b^2 + 2$,
    $3a^2 - 6ab + 7b^2 - 9$, $5a^2 + 7b^2 + 10$

$$\begin{array}{r} 4a^2 - 3ab + b^2 \\ 12ab - 7b^2 + 2 \\ 3a^2 - 6ab + 7b^2 - 9 \\ \underline{5a^2 + 7b^2 + 10} \\ 12a^2 + 3ab + 8b^2 + 3 \end{array}$$

Write the polynomials vertically so that like terms are grouped vertically, leaving blank spaces where necessary. Add the terms in columns.

25. Subtract $4a - a - a$ from $7a - 3a + a$.

$$\begin{array}{r} 7a^4 - 3a^3 + a \\ \underline{4a^4 - a^3 - a^2 + a} \\ 3a^4 - 2a^3 + a^2 + a \end{array}$$

Write the polynomial being subtracted from above the polynomial being subtracted. Subtract by adding the opposite of the term being subtracted.

Solve:

29. $(5x + 2) - (3x - 5) = (6x - 5) + (-7x + 2)$
    $(5x + 2) + (-3x + 5) = (6x - 5) + (-7x + 2)$

$(5x - 3x) + (2 + 5) = (6x - 7x) + (-5 + 2)$
$ 2x + 7 = -x - 3$
$ 3x + 7 = -3$
$ 3x = -10$
$ x = -\dfrac{10}{3}$

On the left, subtract by adding the opposite. Group like terms on both sides. Add x to both sides. Subtract 7 from both sides. Divide both sides by 3.

C

Add or subtract as indicated:

33. $\left(\frac{1}{5}p + 2q - \frac{1}{2}\right) + \left(\frac{3}{2}p - \frac{3}{4}p - 1\right) - \left(-\frac{1}{2}p - \frac{1}{2}p - \frac{1}{2}\right)$

$\left(\frac{1}{5}p + 2q - \frac{1}{2}\right) + \left(\frac{3}{2}p - \frac{3}{4}p - 1\right) + \left(\frac{1}{2}p + \frac{1}{2}p + \frac{1}{2}\right)$ Rewrite the subtraction as addition:

$\frac{1}{5}p + 2q - \frac{1}{2}$
$\frac{3}{2}p - \frac{3}{4}q - 1$
$\frac{1}{2}p + \frac{1}{2}q + \frac{1}{2}$

Write vertically so that like terms are grouped vertically.

$\frac{2}{10}p + \frac{8}{4}q - \frac{1}{2}$
$\frac{15}{10}p - \frac{3}{4}q - \frac{2}{3}$
$\frac{5}{10}p + \frac{2}{4}q + \frac{1}{2}$

Rewrite using the common denominator of each like term.

$\frac{22}{10}p + \frac{7}{4}q - 1$ Combine like terms.

$= \frac{11}{5}p + \frac{7}{4}q - 1$ Reduce.

37. $x^2 - [2x^2 - (5x^2 + 6x) - x + 8] + 3$
$x^2 - [2x^2 + (-5x^2 - 6x) - x + 8] + 3$ Clear the parentheses by rewriting the subtraction as addition.

$x^2 - [(2x^2 - 5x^2) + (-6x - x) + 8] + 3$ Group like terms within the brackets.

$x^2 - [-3x^2 - 7x + 8] + 3$ Now clear the brackets by rewriting the subtraction as addition.

$(x^2 + 3x^2) + (7x) + (-8 + 3)$ Group like terms and
$4x^2 + 7x - 5$ combine.

41. Add $7x + 2y$ to the difference of $6x - y$ and $2x + 3y$.

$6x - y$
$2x + 3y$
$\overline{4x - 4y}$
$7x + 2y$
$\overline{11x - 2y}$

First find the difference of $6x - y$ and $2x + 3y$. Subtract by adding the opposite.

Now add $7x + 2y$ to the difference.

45

45. $(5x^2 + 2x + 5) - (2x^2 + 4x - 4) = 3x^2 - 4x + 5$
     $(5x^2 + 2x + 5) + (-2x^2 - 4x + 4) = 3x^2 - 4x + 5$ On the left, subtract by adding the opposite.

     $(5x^2 - 2x^2) + (2x - 4x) + (5 + 4) = 3x^2 - 4x + 5$ Group like terms on the left.

     $3x^2 - 2x + 9 = 3x^2 - 4x + 5$ Combine like terms.
     $2x + 9 = 5$
     $2x = -4$
     $x = -2$

The solution set is {-2}.

D

49. The cost of manufacturing n brushes is given by $C = \frac{1}{5}n^2 - 14n + 20$. The cost of marketing the same items is given by $C = 5n^2 - 30n + 50$. Find the formula for the total cost of manufacturing and marketing the brushes. Find the total cost when n = 50.

$\frac{1}{5}n^2 - 14n + 20$ Add the two cost formulas together to find the total cost formula.

$\frac{25}{5}n^2 - 30n + 50$ Rewrite the n² terms with a common denominator.

―――――――――――
$\frac{26}{5}n^2 - 44n + 70$ Add the like terms in columns.

$\frac{26}{5}(50)^2 - 44(50) + 70$ Replace n with 50.

$= \frac{25}{5}(2500) - 44(50) + 70$

$= 13000 - 2200 + 70$
$= 10800 + 70$
$= 10870$

So the formula for the total cost is $C = \frac{26}{5}n^2 - 44n + 70$, and the total cost when n = 50 is $10,870.

53. The volume of one container is given by the formula
$V_1 = 2x^3 - 3x^2 + x + 40$ (where x is the length of one side), and the volume of a second container is $V_2 = x^3 + 3x^2 - 5x + 10$. Find a formula for the combined volume, V, of the two containers. Find the combined volume if x = 2.

$V = V_1 + V_2$	The combined volume, V, is the sum of the two volumes.
$V = (2x^3 - 3x^2 + x + 40)$ $+ (x^3 + 3x^2 - 5x + 10)$	Replace V_1 and V_2.
$V = (2x^3 + x^3) + (-3x^2 + 3x^2) + (x - 5x) + (40 + 10)$	Group like terms.
$V = 3x^3 - 4x + 50$	Combine like terms.
$V = 3(2)^3 - 4(2) + 50$	Now replace x with 2.
$V = 3(8) - 8 + 50$	
$V = 24 - 8 + 50$	
$V = 66$	

The formula for the combined volume is $V = 3x^3 - 4x + 50$; the combined volume when x = 2 is 66.

57. Girl Scout troop #52 can sell $\frac{3}{2}t^2 - 2t + 20$ boxes of cookies in t days. Troop #63 can sell $2t^2 - 4t + 10$ boxes of cookies in t days. Find a formula for the total number of boxes of cookies, T, both can sell in t days. How many boxes will both troops have sold in 2 days?

$T = \left[\frac{3}{2}t^2 - 2t + 20\right] + (2t^2 - 4t + 10)$

$= \left[\frac{3}{2}t^2 + 2t^2\right] + (-2t - 4t) + (20 + 10)$

$= \frac{7}{2}t^2 - 6t + 30$

$T = \frac{7}{2}(2)^2 - 6(2) + 30$	Replace t with 2.
$= \frac{7}{2}(4) - 12 + 30$	
$= 14 - 12 + 30$	
$= 32$	

The formula for the total number of boxes is $T = \frac{7}{2}t^2 - 6t + 30$; the total number of boxes sold in 2 days is 32.

E Maintain Your Skills

Perform the indicated operations:

61. $(128) + (43) + (-219) + (-71)$
   $= (171) + (-219) + (-71)$ Perform the indicated
   $= (-48) + (-71)$ additions from left to right.
   $= -119$

65. $-[(-13) + (42) + (-29)]$
    $= -[(29) + (-29)]$
    $= -[0]$
    $= 0$

Do the indicated additions inside the brackets from left to right.

EXERCISES 2.6   MULTIPLICATION OF POLYNOMIALS

A

Multiply:

1. $2(x + 3)$
   $= 2 \cdot x + 2 \cdot 3$

   $= 2x + 6$

   Distributive Law. The factor 2 is multiplied times each term of the polynomial in parentheses. Simplify.

5. $2(3x + 7)$
   $= 2 \cdot 3x + 2 \cdot 7 = 6x + 14$

   Distributive Law. Multiply the factor 2, times each term of the polynomial in parentheses. Simplify.

9. $y(y^2 + 3y - 8)$
   $= y \cdot y^2 + y \cdot 3y + y(-8)$
   $= y^3 + 3y^2 - 8y$

   Distributive Law. Simplify.

13. $3xy(x^2 - 6xy + y^2)$
    $= 3xy \cdot x^2 + 3xy(-6xy) + 3xy(y^2)$
    $= 3x^3y - 18x^2y^2 + 3xy^3$

    Distributive Law. Simplify.

B

17. $(x - 3)(x + 4)$
    $= (x - 3)x + (x - 3)4$
    $= x^2 - 3x + 4x - 12$

    $= x^2 + x - 12$

    Use the Distributive Law. The factor that is "distributed" is $(x - 3)$. Combine like terms.

21. $(3m - 7t)(t - 4m)$
    $= (3m - 7t) \cdot t + (3m - 7t) \cdot (-4m)$

    $= 3mt - 7t^2 - 12m^2 + 28m$
    $= -12m^2 + 31mt - 7t^2$

    Multiply the factor, $(3m - 7t)$, times each of the terms in the second binomial. Combine like terms.

25. $(x^2 + 1)(2x^2 - 7x + 5)$

$$\begin{array}{r} 2x^2 - 7x + 5 \\ x^2 + 1 \\ \hline 2x^4 - 7x^3 + 5x^2 \\ 2x^2 - 7x + 5 \\ \hline 2x^4 - 7x^3 + 7x^2 - 7x + 5 \end{array}$$

Write the problem vertically.

48

Solve:

29. $a(a + 2) - 3[2a - a(a - 7)] = 4a(a - 1) + 21$

$\quad a^2 + 2a - 3[2a - a^2 + 7a] = 4a^2 - 4a + 21$ Remove parentheses by multiplying the polynomials.

$\quad a^2 + 2a - 3[9a - a^2] = 4a^2 - 4a + 21$ Combine like terms within brackets.

$\quad a^2 + 2a - 27a + 3a^2 = 4a^2 - 4a + 21$ Remove brackets by multiplication.

$\quad\quad\quad\quad 4a^2 - 25a = 4a^2 - 4a + 21$ Combine terms.

$\quad\quad\quad\quad\quad\quad -21a = 21$ Solve for a.

$\quad\quad\quad\quad\quad\quad\quad\quad a = -1$

The solution set is {-1}.

C

33. $(a + b - x)(a - b + x)$

$$\begin{array}{r} a + b - x \\ a - b + x \\ \hline ax + bx - x^2 \\ -ab - b^2 \quad\quad + bx \\ a^2 + ab \quad\quad\quad -ax \\ \hline a^2 \quad\quad -b^2 \quad + 2bx - x^2 \end{array}$$

Write the problem in vertical form and multiply.

37. $(2x + 3y)(x^2 - 3xy + y^2 + 6)$

$= (2x + 3y)(x^2) + (2x + 3y)(-3xy)$ Use the Distributive Law.
$\quad + (2x + 3y)(y^2) + (2x + 3y)(6)$

$= (2x^3 + 3x^2y - 6x^2y - 9xy^2 + 2xy^2 + 3y^3$
$\quad + 12x + 18y$

$= 2x^3 - 3x^2y - 7xy^2 + 12x + 3y^3 + 18y$ Combine like terms.

41. $(x + 1)(x - 2)(x + 3)$

$= [(x + 1)(x - 2)](x + 3)$

$= [(x + 1)\cdot x + (x + 1)\cdot -2](x + 3)$

$= [x^2 + x - 2x - 2](x + 3)$

Perform the indicated multiplication from left to right. First distribute the factor $(x + 1)$ to each term of the second binomial $(x - 2)$.

$= (x^2 - x - 2)(x + 3)$

Combine terms within brackets.

$= (x^2 - x - 2)(x) + (x^2 - x - 2)(3)$

Use the Distributive Law again.

$= x^3 - x^2 - 2x + 3x^2 - 3x - 6$

Multiply the factor $(x - x - 2)$ times each term of the binomial $(x + 3)$.

$= x^3 + 2x^2 - 5x - 6$

Combine like terms.

Solve:

45. $x(x + 3) + (x + 2)(x - 7) = 2x(x + 3) - 12$
    $x^2 + 3x + x^2 - 5x - 14 = 2x^2 + 6x - 12$ Remove parentheses by multiplying the polynomials.
    $2x^2 - 2x - 14 = 2x^2 + 6x - 12$ Combine like terms on the left.
    $-8x = 2$ Solve.
    $x = -\frac{1}{4}$

    The solution set is $\left\{-\frac{1}{4}\right\}$.

49. $3x(4x - 3) - (2x - 1)(3x + 2) = 2x(3x - 4)$
    $12x^2 - 9x - [6x^2 + x - 2] = 6x^2 - 8x$ Remove parentheses by multiplying the polynomials.
    $12x^2 - 9x - 6x^2 - x + 2 = 6x^2 - 8x$ Subtract by adding the opposite.
    $6x^2 - 10x + 2 = 6x^2 - 8x$ Combine like terms on the left.
    $-2x = -2$ Solve.
    $x = 1$

    The solution set is $\{1\}$.

D

53. Oil costing $1.30 per quart is to be mixed with oil costing $1.60 per quart to make 90 quarts of a mixture that would cost $1.42 per quart. How much of each grade should be used?

    <u>Simpler word form:</u>
    $\begin{bmatrix} \text{Quarts of oil} \\ \text{at \$1.30 per} \\ \text{quart} \end{bmatrix} + \begin{bmatrix} \text{Quarts of oil} \\ \text{at \$1.60 per} \\ \text{quart} \end{bmatrix} = \begin{bmatrix} 90 \text{ quarts} \\ \text{at \$1.42} \\ \text{per quart} \end{bmatrix}$

    <u>Select a variable:</u>
    Let x represent the number of quarts of oil costing $1.30 per quart; then 90 - x represents the number of quarts costing $1.60 per quart.

    <u>Translate to algebra:</u>
    $1.30x + 1.60(90 - x) = 90(1.42)$

    <u>Solve:</u>
    $1.30x + 144 - 1.6x = 127.8$ Multiply.
    $-0.3x + 144 = 127.8$
    $-0.3x = -16.2$
    $x = 54$
    And $90 - x = 90 - 54 = 36$

    <u>Answer:</u>
    So, 54 quarts at $1.30 per quart mixed with 36 quarts at $1.60 per quart will make 90 quarts costing $1.42 per quart.

57. Linda jogs to a lake at a rate of 4 miles/hr, then walks back at a rate of 1.5 miles/hr. If it takes her one hour longer to walk back, how long did it take her to jog to the lake?

Simpler word form:

$$\begin{pmatrix} \text{Time taken} \\ \text{to walk} \\ \text{back} \end{pmatrix} - \begin{pmatrix} \text{Time taken} \\ \text{to jog} \\ \text{to lake} \end{pmatrix} = 1 \text{ hour}$$

Select variable:
Let x represent the distance from Linda's starting place to the lake.

	Distance	÷ Rate	= Time
Jogging to lake	x	4	$\frac{x}{4}$
Walking from Lake	x	1.5	$\frac{x}{1.5}$

Translate to algebra:
$\frac{x}{1.5} - \frac{x}{4} = 1$

Solve:

4x - 1.5x = 6   Multiply by the common
   2.5x = 6   denominator which is
      x = 2.4   (1.5)(4) = 6.

So $\frac{x}{4}$, the time it took Linda to jog to the lake, equals $\frac{2.4}{4} = 0.6$ or $\frac{3}{5}$ of an hour.

Answer:
It took Linda $\frac{3}{5}$ of an hour, or 36 minutes, to jog to the lake.

E Maintain Your Skills

Multiply or divide as indicated:

61. (-13)(-5)(-2)
    = (65)(-2) = -130   Multiply from left to right.

65. (-8)(-3x)(4)
    = (24x)(4) = 96x

69. Change $-31°F$ to Celsius measure.

$C = \frac{5}{9}(F - 32)$ Use the formula $C = \frac{5}{9}(F - 32)$.

$C = \frac{5}{9}(-31 - 32)$ Replace F with -31.

$C = \frac{5}{9}(-63)$

$C = -35$

So $-31°F$ is equal to $-35°C$.

EXERCISES 2.7   SPECIAL PRODUCTS OF POLYNOMIALS

A

Multiply:

1. $(x + 3)(x + 4) = x^2 + 4x + 3x + 12$ Use FOIL multiplication.
   $= x^2 + 7x + 12$ Combine like terms.

5. $(b - 1)(b - 4) = b^2 - 4b - b + 4$ Use FOIL multiplication.
   $= b^2 - 5b + 4$ Combine like terms.

9. $(x + 6)(x - 7) = x^2 - 7x + 6x - 42$ Use FOIL multiplication.
   $= x^2 - x - 42$ Combine like terms.

13. $(x - 9)^2 = (x)^2 + 2(x)(-9) + (-9)^2$ Square the first term, double the product of the first and last terms, and square the last term.
    $= x^2 - 18x + 81$

B

17. $(3x - 5)(x - 6) = 3x^2 - 18x - 5x + 30$ Use FOIL multiplication.
    $= 3x^2 - 23x + 30$ Combine like terms.

21. $(3x + 7)(5x - 6) = 15x^2 - 18x + 35x - 42$ Use FOIL multiplication.
    $= 15x^2 + 17x - 42$ Combine like terms.

25. $(2a + 5)^2 = (2a)^2 + 2(2a)(5) + (5)^2$ To square the binomial, square the first term, double the product of the first and last term, and square the last term.
    $= 4a^2 + 20a + 25$ Combine like terms.

Solve:

29. $(x + 3)(x + 5) = (x + 7)(x - 5) - 4$
    $x^2 + 5x + 3x + 15 = x^2 - 5x + 7x - 35 - 4$ Use FOIL multiplication to do the indicated multiplications.
    $x^2 + 8x + 15 = x^2 + 2x - 39$ Combine like terms.
    $6x = -54$ Solve for x.
    $x = -9$

The solution set is $\{-9\}$.

C

Multiply:

33. $(5x - 11)(3x + 9) = 15x^2 + 45x - 33x - 99$
    Use FOIL multiplication.
    $= 15x^2 + 12x - 99$ Combine like terms.

37. $\left[\frac{1}{2}x + \frac{1}{3}\right]\left[\frac{1}{3}x + \frac{1}{2}\right] = \frac{1}{6}x^2 + \frac{1}{4}x + \frac{1}{9}x + \frac{1}{6}$
    $= \frac{1}{6}x^2 + \frac{13}{36}x + \frac{1}{6}$ Combine like terms.

41. $[(2x + y) - 5]^2$
    $= (2x + y)^2 - 2 \cdot 5(2x + y) + (-5)^2$ Square the first term, $2x + y$, double the product of the first and last terms, $-5(2x + y)$, and square the last term, (-5).
    $= [(2x)^2 + 2(2xy) + (y)^2] - 10(2x + y) + 25$
    $= 4x^2 + 4xy + y^2 - 20x - 10y + 25$

Solve:

45. $(x - 7)^2 = (x + 6)^2 - 13$
    $x^2 - 14x + 49 = x^2 + 12x + 36 - 13$ Square the binomials.
    $x^2 - 14x + 49 = x^2 + 12x + 23$ Combine like terms on the right.
    $-26x = -26$ Solve for x.
    $x = 1$

The solution set is $\{1\}$.

D

49. A square building lot had to be increased by 5 feet on each side to meet city code. If this increase added 525 square feet to the area, what was the length of a side of the original lot?

Simpler word form:
Original lot area + 525 = New lot area.

Select variable:
Let x represent the length of the original lot. Then x + 5 represents the length of the new enlarged lot.

Translate to algebra:
$x^2 + 525 = (x + 5)^2$

Solve:
$x^2 + 525 = x^2 + 10x + 25$
$-10x = -500$
$x = 50$

Answer:
The length of a side of the original lot was 50 feet.

53. The length of a vegetable garden plot is to be increased by 2 ft while the width is increased by 4 ft. These changes will increase the area by 100 sq ft. If the length was originally 5 ft more than the width, find the new dimensions of the garden plot.

Simpler word form:
$\begin{bmatrix} \text{New} \\ \text{area} \end{bmatrix} - \begin{bmatrix} \text{Old} \\ \text{area} \end{bmatrix} = 100$

Select variable:
Let x represent the original width; then the original length is x + 5.

Translate to algebra:
$(x + 4)[(x + 5) + 2] - x(x + 5) = 100$

Solve:
$(x + 4)(x + 7) - x(x + 5) = 100$
$x^2 + 11x + 28 - x^2 - 5x = 100$
$6x + 28 = 100$
$6x = 72$
$x = 12$
And $x + 5 = 12 + 5 = 17$

The original width, 12, increased by 4, is 16.
The original length, 17, increased by 2, is 19.

Answer:
So, the new dimensions are 19 ft by 16 ft.

57. Johnny is six years older than his brother, Jason. Two years ago, the product of their ages was 20 less than the product of their present ages. How old are the boys now?

Simpler word form:
$$\begin{bmatrix}\text{Johnny's age}\\ \text{2 years ago}\end{bmatrix}\begin{bmatrix}\text{Jason's age}\\ \text{2 years ago}\end{bmatrix} = \begin{bmatrix}\text{Johnny's}\\ \text{age now}\end{bmatrix}\begin{bmatrix}\text{Jason's}\\ \text{age now}\end{bmatrix} - 20$$

Select variable:
Let x represent Jason's age now. Then x + 6 represents Johnny's age now.

	Age now	Age two years ago
Jason	x	x - 2
Johnny	x + 6	x + 4

Make a table to organize the information.

Translate to algebra:
(x + 4)(x - 2) = x(x + 6) - 20

Solve:
x^2 + 2x - 8 = x^2 + 6x - 20
       -4x = -12
         x = 3   Jason's age now.
And x + 6 = 3 + 6 = 9   Johnny's age now.

Answer:
Johnny's age now is 9, and Jason's age now is 3.

E Maintain Your Skills

Perform the indicated operations:

61. 14(23x - 10) - 12(5x - 11)
   = 322x - 140 - 60x + 132   Distributive Law.
   = 262x - 8

65. 15(3 + 14) - 7(11(18 - 23)] + 16
   = 15(17) - 7[11(-5)] + 16
   = 255 - 7[-55] + 16
   = 255 + 385 + 16
   = 640 + 16
   = 656

EXERCISES 2.8   FACTORING POLYNOMIALS: COMMON FACTORS AND
                FACTORING BY GROUPING

A

Factor:

1. $12m - 12n$
    $= 12(m - n)$ The number 12 is a common
                                               factor.

5. $6ab - 12ac + 18ad$
    $= 6a(b - 2c + 3d)$ The largest common numerical
                                               factor is 6. The largest
                                               common variable factor is a.
                                               So the greatest common factor
                                               is 6a.

9. $3xy + 6y + xz + 2z$
    $= (3xy + 6y) + (xz + 2z)$ The four terms do not have
                                               any common factor except 1.
                                               Group the terms in pairs that
                                               have common factors.
    $= 3y(x + 2) + z(x + 2)$ Factor each group.
    $= (3y + z)(x + 2)$ The common factor of the new
                                               expression is the binomial
                                               $(x + 2)$.

Solve:

13. $(x - 5)(x - 9) = 0$
    $x - 5 = 0$ or $x - 9 = 0$ Zero-product law. Set each
                                               factor equal to 0.
    $x = 5$ $x = 9$

    The solution set is $\{5, 9\}$.

B

Factor:

17. $18x^2y^2 - 30xy^2$
    $= 6xy^2(3x - 5)$ The largest common numerical
                                               factor is 6. The largest
                                               common variable factor is
                                               xy^2. So the greatest common
                                               factor is $6xy^2$.

21. $20x^2 - 15x + 8xy - 6y$
    $= (20x^2 - 15x) + (8xy - 6y)$

   $= 5x(4x - 3) + 2y(4x - 3)$
   $= (5x + 2y)(4x - 3)$

The four terms have no common factors except 1. Group the terms in pairs that have common factors.
Factor each group.
The common factor of the new expression is the binomial $(4x - 3)$.

25. $2x^2y - 6x^2 + y - 3$
    $= (2x^2y - 6x^2) + (y - 3)$

    $= 2x^2(y - 3) + 1(y - 3)$
    $= (2x^2 + 1)(y - 3)$

The four terms have no common factor except 1. Group the terms in pairs that have common factors.
Factor each group.
The common factor of the new expression is $(y - 3)$.

Solve:

29. $20x^2 + 25x + 12x + 15 = 0$
    $5x(4x + 5) + 3(4x + 5) = 0$

    $(5x + 3)(4x + 5) = 0$
    $5x + 3 = 0$ or $4x + 5 = 0$
    $x = -\frac{3}{5}$ or $x = -\frac{5}{4}$

    The solution set is $\left\{-\frac{3}{5}, -\frac{5}{4}\right\}$.

Factor by grouping pairs of terms with common factors.
Factor the left side.
Zero-product property.

C

Factor:

33. $6x^3y^2z - 9x^2y^3z - 24x^2y^2z^2$
    $= 3x^2y^2z(2x - 3y - 8z)$

The greatest common factor is $3x^2y^2z$.

37. $42x^2 + 35xy - 18xy - 15y^2$
    $= (42x^2 + 35xy) + (-18xy - 15y^2)$
    $= 7x(6x + 5y) - 3y(6x + 5y)$
    $= (7x - 3y)(6x + 5y)$

Factor by grouping.
Factor each pair.

Solve:

41. $x^2 + 3x - 5x - 15 = 0$
    $(x^2 + 3x) + (-5x - 15) = 0$
    $x(x + 3) - 5(x + 3) = 0$
    $(x - 5)(x + 3) = 0$
    $x - 5 = 0$ or $x + 3 = 0$
    $x = 5$ or $x = -3$

    The solution set is $\{-3, 5\}$.

Factor the left side by grouping.

Zero-product law.

45. $x^2 - ax = 0$
    $x(x - 1) = 0$ Factor the left side.
    $x = 0$ or $x - a = 0$ Zero-product law.
               $x = a$

    The solution set is {0, a}.

D

49. If the average profit (p) per set of dishes sold is given by $p = \frac{1}{4}q^2 - 36q + 42$, where q is the number of sets sold, how many sets must be sold to have an average profit of $42?

    Formula:
    $p = \frac{1}{4}q^2 - 36q + 42$

    Substitute:
    $42 = \frac{1}{4}q^2 - 36q + 42$

    Solve:
    $0 = \frac{1}{4}q^2 - 36q$
    $0 = q\left[\frac{1}{4}q - 36\right]$
    $q = 0$ or $\frac{1}{4}q - 36 = 0$ Reject $q = 0$. There is no
                   $\frac{1}{4}q = 36$ average profit if no sets
                     $q = 144$ are sold.

    Answer:
    The number of sets that must be sold to have an average profit of $42 is 144.

53. Given the equation $S = 16t - 4t^2$, where s is the height of a falling object and t is the time the object falls, how long does it take an object to hit the ground? (S = 0)

    Formula:
    $S = 16t - 4t^2$

    Substitute:
    $0 = 16t - 4t^2$
    $0 = 4t(4 - t)$
    $4t = 0$ or $4 - t = 0$
    $t = 0$ or $-t = -4$ Reject $t = 0$.
                  $t = 4$

    Answer:
    It takes the object 4 seconds to hit the ground.

57. The cost of manufacturing x toy turtles is
C = 2x² - 1600x + 5000. How many turtles can be produced at a cost of $5000?

   Formula:
   C = 2x² - 1600x + 5000

   Substitute:
   5000 = 2x² - 1600x + 5000   Substitute C = 5000.

   Solve:
   0 = 2x² - 1600x
   0 = 2x(x - 800)
   2x = 0 or x - 800 = 0
   x = 0 or   x = 800   Reject x = 0, since there is
                                      no cost if no toys are
                                      produced.

   Answer:
   So 800 toys can be produced at a cost of $5000.

E Maintain Your Skills

Solve:

61. (x - 3) - (4x - 5) - (3 - 2x) = 7
           x - 3 - 4x + 5 - 3 + 2x = 7
                            -x - 1 = 7
                                -x = 8
                                 x = -8

   The solution set is {-8}.

65. 16(35x + 8) = 33(5x - 44)
       560x + 128 = 165x - 1452
             395x = -1580
                x = -4

   The solution set is {-4}.

59

EXERCISES 2.9   FACTORING POLYNOMIALS: TRINOMIALS

A

Factor:

1. $x^2 + 12x + 35$
    mn = 35   m + n = 12
    (1)(35)   36
    (5)(7)   12

    $(x + 5)(x + 7)$

    Find two numbers whose product is 35 (mn = 35) and whose sum is 12 (m + n = 12). None of the negative pairs of factors are listed, since the middle term is positive.

5. $a^2 - 7a - 18$
    mn = -18   m + n = -7
    (1)(-18)   -17
    (2)(-9)   -7

    $(a + 2)(a - 9)$

    Find two numbers whose product is -18 (mn = -18) and whose sum is -7 (m + n = -7).

9. $a^2 - 3a - 10$
    Try $(a + 1)(a - 10)$.
    The product is $a^2 - 9a - 10$.

    Try $(a + 2)(a - 5)$.
    The product is $a^2 - 3a - 10$, so the factors are $(a + 2)(a - 5)$.

    Factor by trial and error.

Solve:

13. $x^2 - 3x + 2 = 0$
    $(x - 1)(x - 2) = 0$
    $x - 1 = 0$ or $x - 2 = 0$
    $x = 1$ or $x = 2$

    The solution set is {1, 2}.

    Factor the left side.
    Zero-product law.

B

17. $x^2 + x - 110$
    mn = -110   m + n = 1
    (-1)(110)   109
    (-2)(55)   53
    (-5)(22)   17
    (-10)(11)   1

    List possible combinations of m and n.

    The last pair is the correct combination.

    $(x - 10)(x + 11)$

60

21. $2a^2 + 9a - 18$

    $mn = (2)(-18) = -36$ $m + n = 9$ List possible combinations
    $(-1)(36)$ 35   of m and n.
    $(-2)(18)$ 16
    $(-3)(12)$ 9

    $2a^2 - 3a + 12a - 18$ Rewrite the middle term:
                                         $9a = -3a + 12a$.
    $(2a^2 - 3a) + (12a - 18)$ Factor by grouping.
    $a(2a - 3) + 6(2a - 3)$
    $(a + 6)(2a - 3)$

Solve:

25. $x^2 - 6x - 72 = 0$
    $(x - 12)(x + 6) = 0$ Factor the left side.
    $x - 12 = 0$ or $x + 6 = 0$ Zero-product law.
        $x = 12$ or $x = -6$

    The solution set is $\{-6, 12\}$.

29. $3b^2 - b - 2 = 0$
    $(3b + 2)(b - 1) = 0$ Factor the left side.
    $3b + 2 = 0$ or $b - 1 = 0$ Zero-product law.
        $3b = -2$ or $b = 1$
         $b = -\frac{2}{3}$

    The solution set is $\left\{-\frac{2}{3}, 1\right\}$.

C

Factor:

33. $x^2 - 6x - 216$
    Try $(x + 2)(x - 108)$.   Factor by trial and error.
    The product is $x^2 - 106x - 216$.   The middle term, $-106x$, is
                                           not close to $-6x$, so try
                                           factors of 216 that are
                                           closer together than 2 and
                                           108.

    Try $(x + 9)(x - 24)$.
    The product is $x^2 - 15x - 216$.   The middle term, $-15x$, is
                                           closer to $-6x$, so again try
                                           factors that are closer
                                           together than 9 and 24.

    Try $(x + 12)(x - 18)$.
    The product is $x^2 - 6x - 216$,
    so the factors are $(x + 12)(x - 18)$.

37. $6y^2 + 37y + 55$

    $mn = (6)(55) = 330 \quad m + n = 37$

(1)(330)	331
(2)(165)	167
(3)(110)	113
(5)(66)	71
(6)(55)	61
(10)(33)	43
(11)(30)	41
(15)(22)	37

List possible combinations of m and n.

The last pair is the correct combination.

$6y^2 + 15y + 22y + 55$

$(6y^2 + 15y) + (22y + 55)$
$3y(2y + 5) + 11(2y + 5)$
$(3y + 11)(2y + 5)$

Rewrite the middle term:
$37y = 15y + 22y$
Factor by grouping.

41. $(x + 2y)^2 - (x + 2y) - 42$
Let $x + 2y = p$

$p^2 - p - 42$

Since the polynomial has two expressions in $(x + 2y)$, the factors will be easier to see using substitution.
Factor by trial and error.
The middle term, $-1p$, is small, so the factors of 42 should be close.

Try $(p - 7)(p + 6)$.
The product is $p^2 - p - 42$, so the factors are $(p - 7)(p + 6)$.
$(x + 2y - 7)(x + 2y + 6)$

Finally, replace p by $x + 2y$.

45. $(6t - 1)^2 - 17(6t - 1) - 38$
Let $p = 6t - 1$.

Since the polynomial has two expressions in $(6t - 1)$, the factor will be easier to see using substitution.

$p^2 - 17p - 38$
$mn = -38 \quad m + n = -17$

(1)(-38)	-37
(2)(-19)	-17

List possible combinations of m and n.

$(p + 2)(p - 19)$
$(6t - 1 + 2)(6t - 1 - 19)$
$(6t + 1)(6t - 20)$

    $(6t + 1)(2)(3t - 10)$
or $2(6t + 1)(3t - 10)$

Finally replace p by $6t - 1$.
Combine like terms within parentheses.
Notice that the terms of the second binomial have a common factor, 2.

Solve:

49. $10x^2 - 31x - 63 = 0$
$(2x - 9)(5x + 7) = 0$ Factor the left side.
$2x - 9 = 0$ or $5x + 7 = 0$ Zero-product law.
    $2x = 9$ or $5x = -7$
      $x = \frac{9}{2}$ or $x = -\frac{7}{5}$

The solution set is $\left\{\frac{9}{2}, -\frac{7}{5}\right\}$.

D

53. The perimeter of a rectangular frame is 36 inches. If the area enclosed by the frame is 72 square inches, find the dimensions of the frame.

Formula:
$\ell w = A$ Formula for the area of a rectangle.

Select variable:
Let x represent the width of the frame. Since half the perimeter or 18 inches is the length plus the width, $18 - x$ represents the length.

Substitute:
$(18 - x)x = 72$ Replace ℓ with $18 - x$, and w with x. Replace A with 72.

Solve:
$18x - x^2 = 72$
        $0 = x^2 - 18x + 72$
        $0 = (x - 6)(x - 12)$ Factor.
$x - 6 = 0$ or $x - 12 = 0$ Zero-product law.
    $x = 6$ or $x = 12$

Answer:
The dimensions of the frame are 12 in. by 6 in.

57. A strip of metal 30 inches wide is to be bent into a rectangular conduit that is enclosed on all four sides. If the area of the cross section is 50 sq in., what are the dimensions of the cross section?

Simpler word form:
$\begin{bmatrix}\text{Area of}\\ \text{cross}\\ \text{section}\end{bmatrix} = \begin{bmatrix}\text{Width of}\\ \text{cross}\\ \text{section}\end{bmatrix}\begin{bmatrix}\text{Length of}\\ \text{cross}\\ \text{section}\end{bmatrix}$

Select variable:
Let x represent the width of the cross section. Since half the perimeter, 15, is the sum of the width and the length, then $15 - x$ represents the length.

Translate to algebra:
$50 = x(15 - x)$

Solve:
$$50 = 15x - x^2$$
$$x^2 - 15x + 50 = 0$$
$$(x - 10)(x - 5) = 0$$
$$x - 10 = 0 \text{ or } x - 5 = 0$$
$$x = 10 \text{ or } x = 5$$

If $x = 10$, then the length, $15 - x = 15 - 10 = 5$.
If $x = 5$, then the length, $15 - x = 15 - 5 = 10$.

Answer:
The dimensions of the cross section are 5 in. by 10 in.

E   Maintain Your Skills

Solve for x:

61. $7x - 6x = 3x - 2x + 4$
           $x = x + 4$
           $0 = 4$ Contradiction.

   The solution set is the null set, ∅.

65. $t = \frac{3}{5}x - 9$

           $t + 9 = \frac{3}{5}x$

   $\frac{5}{3}(t + 9) = x$

           $x = \frac{5(t + 9)}{3} = \frac{5t + 45}{3}$

   or $\frac{5}{3}t + \frac{5}{3}(9) = \frac{5}{3}t + 15$

EXERCISES 2.10   FACTORING POLYNOMIALS: SPECIAL CASES

A

Factor, if possible:

1. $a^2 - 1$
     $(a)^2 - (1)$ The difference of the squares of a and 1.
     $(a + 1)(a - 1)$ The factors are conjugate binomials.

5. $y^2 + 16$ The sum of two squares cannot
     $(y)^2 + (4)^2$ be factored by the methods of
                                      this section.
     $y^2 + 16$ is a prime polynomial.

9. $c^2 + 6c + 9$
   $(c)^2 + 2(3)(c) + (3)^2$

   $(c + 3)^2$

The polynomial begins and ends with squares, c^2 and 9. The middle term is twice the product of the expressions that are squared. The middle term is positive, so the binomial is a sum.

13. $x^3 - 1$
    $(x)^3 - (1)^3$

    $(x - 1)(x^2 + x + 1)$

The difference of two cubes. The factors are a binomial and a trinomial. The binomial is the difference of x and 1. The trinomial is the square of the first term, x, plus the product of the two terms, x·1, plus the square of the last term, 1.

Solve:

17. $x^2 - 49 = 0$
    $(x - 7)(x + 7) = 0$
    $x - 7 = 0$ or $x + 7 = 0$
    $x = 7$ or $x = -7$

    The solution set is {-7, 7}.

Factor the difference of squares on the left. Zero-product law.

B

Factor, if possible:

21. $9a^2b^2 - 64c^2$
    $(3ab)^2 - (8c)^2$
    $(3ab + 8c)(3ab - 8c)$

Difference of squares. The factors are conjugate binomials.

25. $4a^2 - 28a + 49$
    $(2a)^2 - 2(2a)(7) + (7)^2$

    $(2a - 7)^2$

A perfect square trinomial since the first and last terms are squares, and the middle term is twice the product of the expressions that are squared.

Solve:

29. $$4x^2 - 25 = 0$$
$$(2x - 5)(2x + 5) = 0$$

Factor the difference of squares on the left.
Zero-product law.

$$2x - 5 = 0 \text{ or } 2x + 5 = 0$$
$$x = \frac{5}{2} \text{ or } \qquad x = -\frac{5}{2}$$

The solution set is $\left\{-\frac{5}{2}, \frac{5}{2}\right\}$.

33. $$4x^2 - 36x = -81$$
$$4x^2 - 36x + 81 = 0$$
$$(2x - 9)^2 = 0$$

Add 81 to both sides.
Factor the perfect square trinomial on the left.
Zero-product law.

$$2x - 9 = 0 \text{ or } 2x - 9 = 0$$
$$x = \frac{9}{2} \text{ or } \qquad x = \frac{9}{2}$$

The solution set is $\left\{\frac{9}{2}\right\}$.

C

Factor, if possible:

37. $(a + b)^2 - c^2$
$(a + b)^2 - (c)^2$
$(a + b + c)(a + b - c)$

Difference of two squares. The factors are conjugate binomials.

41. $100y^2 - 260y + 169$
$(10y)^2 - 2(10y)(13) + (13)^2$
$(10y - 13)^2$

A perfect square trinomial. Since the middle term is negative, the binomial is a difference.

45. $1000\,c^3d^3 - 343$
$(10\,cd)^3 - (7)^3$

$(10cd - 7)(100c^2d^2 + 70cd + 49)$

Difference of cubes. The factors are a binomial and a trinomial.
The binomial is the difference of 10cd and 7.
The trinomial is the square of 10cd, plus the product of the first and last terms, plus the square of 7.

49. $(a - b)^2 - (x - y)^2$
$[(a - b) + (x - y)][(a - b) - (x - y)]$

$(a - b + x - y)(a - b - x + y)$

Difference of squares. The factors are conjugate binomials. Simplify.

53. $144 - (4t^2 + 4st + s^2)$
$144 - (2t + s)^2$ The trinomial in parentheses is a perfect square.
$(12)^2 - (2t + s)^2$ Difference of squares.
$[12 + (2t + s)][12 - (2t + s)]$ The factors are conjugate binomials.
$(12 + 2t + s)(12 - 2t - s)$ Simplify.

Solve:

57. $4x^2 - 28x + 49 = 0$
$(2x - 7)^2 = 0$ A perfect square trinomial.
$2x - 7 = 0$ or $2x - 7 = 0$ Zero-product property.
$x = \frac{7}{2}$ or $x = \frac{7}{2}$ A double root.

The solution set is $\left\{\frac{7}{2}\right\}$.

D

61. The square of Elsa's age in 12 years minus 24 times her age in 24 years is nine. How old is Elsa today?

Simpler word form:
$\left[\begin{array}{c}\text{Elsa's age} \\ \text{in 12 years}\end{array}\right]^2 - 24\left[\begin{array}{c}\text{her age} \\ \text{in 24 years}\end{array}\right] = 9$

Select variable:
Let x represent Elsa's age now. Then x + 12 represents her age in 12 years, and x + 24 represents her age in 24 years.

Translate to algebra:
$(x + 12)^2 - 24(x + 24) = 9$

Solve:
$x^2 + 24x + 144 - 24x - 576 - 9 = 0$
$x^2 - 441 = 0$ Difference of squares.
$(x + 21)(x - 21) = 0$ Zero-product law.
$x = -21$ or $x = 21$ Reject -21 since negative numbers do not represent age.

Answer:
Elsa is 21 years old.

65. Check the formula $a^3 + b^3 = (a + b)(a^2 - ab + b^2)$ by multiplying the polynomials on the right.

$$\begin{array}{r} a^2 - ab + b^2 \\ \underline{a + b} \\ a^3 - a^2b + ab^2 \\ \underline{ a^2b - ab^2 + b^3} \\ a^3 + b^3 \end{array}$$

True, $a^3 + b^3 = (a + b)(a^2 - ab + b^2)$.

67

E Maintain Your Skills

Solve:

69. $2y - [y - (5 - y) + y] = -8$
     $2y - [y - 5 + y + 6] = -8$ Perform indicated operations
     $2y - [2y + 1] = -8$ on the left.
     $2y - 2y - 1 = -8$
     $-1 = -8$ Contradiction.

The solution set is the null set, ∅.

73. $3x - 2 - (2x + 3) \geq 11$
     $3x - 2 - 2x - 3 \geq 11$
     $x - 5 \geq 11$
     $x \geq 16$

The solution set is $\{x | x \geq 16\}$.

EXERCISES 2.11   FACTORING POLYNOMIALS: A REVIEW

A

Factor, if possible:

1. $9c + 6b - 15$ Look for a greatest common
     $3(3c + 2b - 5)$ factor first. The GCF is 3.
                                          The resulting trinomial
                                          cannot be further factored.

5. $x^2 - 8x + 12$
     $mn = 12$ $m + n = -8$ List possible combinations
     $(1)(-12)$ -13 of m and n.
     $(-2)(-6)$ -8 No more combinations are
                                          needed.
     $(x - 2)(x - 6)$

9. $a^2 + 81$ The sum of two squares cannot
     $(a)^2 + (9)^2$ be factored by the methods in
                                          this section.
     $a^2 + 81$ is a prime polynomial.

Solve:

13. $4x^2 - 8x = 0$
     $4x(x - 2) = 0$ The GCF is 4x.
     $4x = 0$ or $x - 2 = 0$ Zero-product law.
     $x = 0$ or $x = 2$

The solution set is $\{0, 2\}$.

17. $x^2 - 8x + 15 = 0$

$(x - 3)(x - 5) = 0$
$x - 3 = 0$ or $x - 5 = 0$
$x = 3$ or $x = 5$

The solution set is $\{3, 5\}$.

There are only two combinations of factors for 15: $1 \cdot 15$ and $3 \cdot 5$.
Trial and error.
Zero-product law.

B

Factor, if possible:

21. $x^2 - 16x + 55$

$(x - 11)(x - 5)$

The two combinations of factors of 55 are: $1 \cdot 55$ and $5 \cdot 11$.

25. $3x^2 + 9x - 30$
$3(x^2 + 3x - 10)$
$3(x + 5)(x - 2)$

The GCF is 3.

29. $25x^2 + 1$
$(5x)^2 + (1)^2$

$25x^2 + 1$ is a prime polynomial.

The sum of squares cannot be factored by the methods of this section.

33. $1000c^3 - d^3$
$(10c)^3 - (d)^3$
$(10c - d)(100c^2 + 10cd + d^2)$

Difference of cubes.

37. $2x^2 + 13x + 21$
$(2x + 1)(x + 21)$
$(2x + 21)(x + 1)$
$(2x + 3)(x + 7)$
$(2x + 7)(x + 3)$

$2x^2 + 13x + 21 = (2x + 7)(x + 3)$

There are two combinations of factors of 21: $1 \cdot 21$ and $3 \cdot 7$.

Trial and error.

Solve:

41. $x^2 - 11x + 24 = 0$
$(x - 3)(x - 8) = 0$

$x - 3 = 0$ or $x - 8 = 0$
$x = 3$ or $x = 8$

The solution set is $\{3, 8\}$.

There is no common factor. For 24, there are four combinations of factors: $1 \cdot 24$, $2 \cdot 12$, $3 \cdot 8$, and $4 \cdot 6$.
Zero-product law.

45. $\quad 6x^2 + 6x - 12 = 0$
  $6(x^2 + x - 2) = 0$
  $6(x + 2)(x - 1) = 0$
  $(x + 2)(x - 1) = 0$
  $x + 2 = 0$ or $x - 1 = 0$
  $\quad x = -2 \quad\quad\quad x = 1$

  The GCF is 6.
  Trial and error.
  Divide both sides by 6.
  Zero-product law.

  The solution set is $\{-2, 1\}$.

C

Factor completely:

49. $36w^2 + 60w + 25$

  $mn = 900 \quad m + n = 60$
  $10 \cdot 90 \quad\quad\quad 100$
  $12 \cdot 75 \quad\quad\quad\; 87$
  $20 \cdot 45 \quad\quad\quad\; 64$
  $30 \cdot 30 \quad\quad\quad\; 60$

  There is no common factor. Use an m, n chart. Use only positive factors since both the sum, 60, and product, 900, are positive.

  $36w^2 + 30w + 30w + 25$
  $6w(6w + 5) + 5(6w + 5)$
  $(6w + 5)(6w + 5)$ or $(6w + 5)^2$

  Rewrite $60w$ as $30w + 30w$.

53. $250x^3 - 16$
  $2(125x^3 - 8)$
  $2[(5x)^3 - (2)^3]$
  $2(5x - 2)(25x^2 + 10x + 4)$

  The GCF is 2.
  Difference of cubes.

57. $28a^3 + 58a^2 - 30a$
  $2a(14a^2 + 29a - 15)$
  $2a(2a + 5)(7a - 3)$

  The GCF is $2a$.
  Factor by trial and error.

61. $x^4 - x^2 - 12$
  $(x^2 - 4)(x^2 + 3)$

  Trial and error. Note that the factors of x^4 are $x^2 \cdot x^2$. Also the first binomial is the difference of squares.

  $(x + 2)(x - 2)(x^2 + 3)$

65. $x^6 - 1$
  $(x^3 + 1)(x^3 - 1)$
  $(x + 1)(x^2 - x + 1)(x - 1)(x^2 + x + 1)$

  Difference of squares. Sum of cubes, and difference of cubes.

Solve:

69. $6x^2 - 41x + 70 = 0$
    $m \cdot n = 420 \quad m + n = -41$ Use an m, n chart.
    $-42 \cdot -10 \qquad\qquad -52$
    $-21 \cdot -20 \qquad\qquad -41$
        $6x^2 - 21x - 20x + 70 = 0$ Rewrite $-41x$ as $-21x - 20x$.
    $3x(2x - 7) - 10(2x - 7) = 0$ Factor by grouping.
           $(3x - 10)(2x - 7) = 0$
    $3x - 10 = 0$ or $2x - 7 = 0$ Zero-product law.
       $x = \dfrac{10}{3}$ or $x = \dfrac{7}{2}$

    The solution set is $\left\{\dfrac{10}{3}, \dfrac{7}{2}\right\}$.

73. $96w^2 + 1 = 28w$
    $96w^2 - 28w + 1 = 0$ Subtract $28w$ from both sides.
    $(4w - 1)(24w - 1) = 0$ Factor.
    $4w - 1 = 0$ or $24w - 1 = 0$ Zero-product law.
       $x = \dfrac{1}{4}$ or $w = \dfrac{1}{24}$

    The solution set is $\left\{\dfrac{1}{24}, \dfrac{1}{4}\right\}$.

E   Maintain Your Skills

Perform the indicated operations:

77. $-2(x - 4) - 2[5(x - 1)] - 3(2 - x)$
    $-2(x - 4) - 2(5x - 5) - 3(2 - x)$ Use the distributive law to clear the brackets.
    $-2x + 8 - 10x + 10 - 6 + 3x$ Use the distributive law to clear the parentheses.
    $-9x + 12$ Combine like terms.

Solve:

81. $3[2a - 3(a + 6) - (a + 2)] + 4 = (a - 4) + 2$
        $3[2a - 3a - 18 - a - 2] + 4 = a - 4 + 2$ Perform the
                             $3[-2a - 20] + 4 = a - 2$ indicated opera-
                                 $-6a - 60 + 4 = a - 2$ tions on both
                                     $-6a - 56 = a - 2$ sides.
                                        $-7a - 56 = -2$ Subtract a from both sides.
                                             $-7a = 54$ Add 56 to both sides.
                                              $a = -\dfrac{54}{7}$ Divide both sides by 7.

    The solution set is $\left\{-\dfrac{54}{7}\right\}$.

85. Two planes leave Tempe traveling in opposite directions. If their average speeds are 420 mph and 525 mph, how long will it be until they are 1260 miles apart?

Simpler word form:

$$\begin{bmatrix} \text{Distance traveled} \\ \text{by one plane} \end{bmatrix} + \begin{bmatrix} \text{Distance traveled} \\ \text{by other plane} \end{bmatrix} = 1260 \text{ mi}$$

Select variable:
Let t represent the time traveled, then 420t represents the distance traveled by one plane and 525t represents the distance traveled by the other. (Distance equals rate times time.)

Translate to algebra:
420t + 525t = 1260

Solve:
$$945t = 1260$$
$$t = 1\frac{1}{3}$$

Answwer:

The time it takes for the two planes to be 1260 mi apart is $1\frac{1}{3}$ hours, or 1 hour and 20 minutes.

CHAPTER 3

RATIONAL EXPRESSIONS

EXERCISES 3.1   PROPERTIES OF RATIONAL EXPRESSIONS

A

Build these rational expressions by finding the missing numerator. You do not need to state the variable restrictions:

1. $\dfrac{m}{x} = \dfrac{?}{x^2}$

   $\dfrac{m}{x} = \dfrac{m(x)}{x(x)} = \dfrac{mx}{x^2}$

   The missing numerator is mx.

   To find the missing numerator find the factor that was introduced into the denominator (x). This can be done by division ($x^2 \div x$). Now introduce this factor (x) into the numerator.

5. $\dfrac{10}{2w + 7} = \dfrac{?}{2w^2 + 7w}$

   $\dfrac{10}{2w + 7} = \dfrac{10(w)}{(2w + 7)(w)}$
   $= \dfrac{10w}{2w^2 + 7w}$

   The missing numerator is 10w.

   To find the missing numerator find the factor that was introduced into the denominator by factoring $2w^2 + 7w$.
   The factors of $2w^2 + 7w$ are w and 2w + 7.
   Since w was introduced as a factor in the denominator, the Basic Principle of Fractions calls for it to be introduced in the numerator as well.

Reduce:

9. $\dfrac{3abc}{18abx} = \dfrac{(3ab)c}{(3ab)6x}$

   $= \dfrac{\cancel{(3ab)}^{1} c}{\cancel{(3ab)}_{1} 6x} = \dfrac{c}{6x}$

   Identify the common factors, 3ab, in the numerator and denominator.

   Reduce by dividing out the common factor.

13. $\dfrac{5(y - 2)}{6(y - 2)} = \dfrac{5\cancel{(y - 2)}^{1}}{6\cancel{(y - 2)}_{1}}$

    $= \dfrac{5}{6}$

    Reduce by dividing out the common factor, y − 2.

B

Build these rational expressions by finding the missing numerator. You do not need to state the variable restrictions:

17. $\dfrac{-3}{6mn} = \dfrac{?}{-12m^2n}$

$\dfrac{-3}{6mn} = \dfrac{-3(-2m)}{6mn(-2m)}$

$= \dfrac{6m}{-12m^2n}$

Divide $-12m^2n$ by $6mn$ to find the factors that were introduced into the denominator.
$-12m^2n \div 6mn = -2m$
Introduce $-2m$ as a factor into the numerator.

The missing numerator is $6m$.

21. $\dfrac{3}{x-1} = \dfrac{?}{x^2-1}$

$\dfrac{3}{x-1} = \dfrac{3(x+1)}{(x-1)(x+1)}$

$= \dfrac{3x+3}{x^2-1}$

Find the factor that was introduced into the denominator by factoring $x^2 - 1$. The factors of $x^2 - 1$ are $x + 1$ and $x - 1$.
Since $(x + 1)$ was introduced into the denominator, it must be introduced into the numerator as well.

The missing numerator is $3x + 3$.

Reduce. You need not state the variable restrictions.

25. $\dfrac{54a^2b^2c^3}{72a^4bc^2}$

$= \dfrac{\cancel{18a^2bc^2}^{\,1}(3bc)}{\cancel{18a^2bc^2}_{\,1}(4a^2)} = \dfrac{3bc}{4a^2}$

Reduce by dividing out the common factors, $18a^2bc^2$, in the numerator and denominator.

29. $\dfrac{x^2-1}{4x+4} = \dfrac{(\cancel{x+1})(x-1)}{4(\cancel{x+1})}$

$= \dfrac{x-1}{4}$

Factor both numerator and denominator.

Reduce by dividing out the common factor, $(x + 1)$.

C

Build these rational expressions. State the restrictions on the variables so that the denominator is not zero.

33. $\dfrac{m + 2}{2m + 7} = \dfrac{?}{4m^2 - 49}$

$m \neq \pm \dfrac{7}{2}$

The expression $4m^2 - 49 = (2m + 7)(2m - 7)$ and will equal zero if either factor is zero.

$\dfrac{m + 2}{2m + 7} = \dfrac{(m + 2)(2m - 7)}{(2m + 7)(2m - 7)}$, $m \neq \pm \dfrac{7}{2}$

Since $2m - 7$ was introduced as a factor in the denominator, introduce it as a factor in the numerator as well.

$= \dfrac{2m^2 - 3m - 14}{4m^2 - 49}$, $m \neq \pm \dfrac{7}{2}$

Multiply.

The missing numerator is $2m^2 - 3m - 14$, $m \neq \pm \dfrac{7}{2}$.

37. $\dfrac{7x + 2}{2x - 1} = \dfrac{?}{6x^2 - 5x + 1}$

$x \neq \dfrac{1}{2}$, $x \neq \dfrac{1}{3}$

The expression $6x^2 - 5x + 1 = (2x - 1)(3x - 1)$ and will equal zero if either factor is zero.

$\dfrac{7x + 2}{2x - 1} = \dfrac{(7x + 2)(3x - 1)}{(2x - 1)(3x - 1)}$, $x \neq \dfrac{1}{2}, \dfrac{1}{3}$

Introduce $(3x - 1)$ into the numerator.

$= \dfrac{21x^2 - x - 2}{6x^2 - 5x + 1}$, $x \neq \dfrac{1}{2}, \dfrac{1}{3}$

Multiply.

The missing numerator is $21x^2 - x - 2$, $x \neq \dfrac{1}{2}, \dfrac{1}{3}$.

Reduce. You need not state the variable restrictions:

41. $\dfrac{x^2 - 16}{x^2 - x - 12} = \dfrac{(x + 4)(x - 4)}{(x + 3)(x - 4)}$

Identify the common factor $(x - 4)$ in the numerator and denominator.

$= \dfrac{x + 4}{x + 3}$

Reduce by dividing out thhe common factor.

45. $\dfrac{2x^2 - x - 15}{2x^2 + 7x + 15} = \dfrac{(2x + 5)(x - 3)}{(2x + 5)(x + 1)}$

Identify the common factor $(2x + 5)$ in the numerator and denominator.

$= \dfrac{x - 3}{x + 1}$

Reduce by dividing out the common factor.

49. $\dfrac{x^2 - 2x - 3}{3x^3 + 3} = \dfrac{\cancel{(x+1)}(x-3)}{3\cancel{(x+1)}(x^2 - x + 1)}$ Factor. Divide out the common factor, $(x+1)$.

$\phantom{\dfrac{x^2 - 2x - 3}{3x^3 + 3}} = \dfrac{x - 3}{3(x^2 - x + 1)}$

E Maintain Your Skills

Add or subtract as indicated:

53. Find the sum of $21t - 5y - 13c + 6$ and $-18t + 22y - 14c - 16$.

$\begin{array}{r} 21t - 5y - 13c + 6 \\ \underline{-18t + 22y - 14c - 16} \\ 3t + 17y - 27c - 10 \end{array}$

57. What is the degree of the polynomial $x^2 - 3xy - 4y$?

Degree 2   The highest degree of any of the terms is 2.

61. A new carpet was installed in a square room. A six-inch border of hardwood floor was left around the perimeter of the room. If the uncarpeted area is 41 square feet, what is the length of a side of the room?

<u>Simpler word form:</u>
$\begin{bmatrix} \text{Floor} \\ \text{area} \end{bmatrix} - \begin{bmatrix} \text{Area of} \\ \text{rug} \end{bmatrix} = 41 \text{ft}^2$

<u>Select a variable:</u>
Let s represent the length of the side of the room. Since the six-inch border is on both sides of the length of the room, the length of the rug is 12 in. or 1 ft less than s, or $s - 1$.

<u>Translate to algebra:</u>
$s^2 - (s - 1)^2 = 41$ $A = s^2$ is the formula for the area of a square.

<u>Solve:</u>
$s^2 - (s^2 - 2s + 1) = 41$
$s^2 - s^2 + 2s - 1 = 41$
$2s - 1 = 41$
$2s = 42$
$s = 21$

<u>Answer:</u>
The length of the sides of the room is 21 ft.

EXERCISES 3.2   MULTIPLICATION AND DIVISION OF RATIONAL EXPRESSIONS

Multiply or divide and simplify. You need not state variable restrictions:

A

1. $\dfrac{y}{8} \cdot \dfrac{16}{w} = \dfrac{16y}{8w}$ Multiply the numerators and denominators.

   $= \dfrac{(8)(2y)}{(8)(w)} = \dfrac{2y}{w}$ Reduce.

   or

   $\dfrac{y}{8} \cdot \dfrac{16}{w} = \dfrac{y}{\underset{1}{8}} \cdot \dfrac{\overset{2}{\cancel{16}}}{w} = \dfrac{2y}{\cancel{w}}$ Cancel common factors before multiplying.

5. $\dfrac{a}{\underset{2}{\cancel{12}}} \cdot \dfrac{\overset{-1}{\cancel{-6}}}{a-3} = \dfrac{-a}{2a-6}$ Cancel the common factor 6, and multiply.

Divide and simplify. State the restrictions on the variables.

9. $6 \div \dfrac{x}{5} = \dfrac{6}{1} \cdot \dfrac{5}{x},\ x \neq 0$ Change division to multiplication by multiplying by the reciprocal of the divisor. Restrict the variables.

    $= \dfrac{30}{x},\ x \neq 0$

13. $(x-4) \div \dfrac{x-3}{2}$

    $= \dfrac{x-4}{1} \cdot \dfrac{2}{x-3},\ x \neq 3$ Change division to multiplication (invert and multily). Restrict the variable.

    $= \dfrac{2(x-4)}{x-3} = \dfrac{2x-8}{x-3},\ x \neq 3$ Multiply. There are no common factors.

B

Multiply or divide and simplify. You need no state the restrictions:

17. $\dfrac{6ax}{5by} \cdot \dfrac{10y}{18x} = \dfrac{60axy}{90bxy}$ Multiply the numerators and denominators.

    $= \dfrac{(\cancel{30xy})(2a)}{(\cancel{30xy})(3b)} = \dfrac{2a}{3b}$ Reduce.

21. $\dfrac{\cancel{7}^{1}}{3(\cancel{x+2})} \cdot \dfrac{\cancel{x+2}}{\underset{2}{\cancel{14}}} = \dfrac{1}{6}$ Reduce and multiply.

Multiply or divide and simplify. State the restrictions on the variables:

25. $\dfrac{3x + 6}{2x - 4} \cdot \dfrac{5x - 10}{6x + 12}$

$= \dfrac{\cancel{3}^{1}(\cancel{x+2})}{2(\cancel{x-2})} \cdot \dfrac{5(\cancel{x-2})}{\underset{2}{\cancel{6}}(\cancel{x+2})},\ x \neq -2,\ 2$ Factor, restrict the variables and reduce.

$= \dfrac{5}{4},\ x \neq \pm 2$ Multiply.

29. $\dfrac{2z - 26}{z^2 - 5z} \div \dfrac{3z - 39}{z^2 + 3z}$

$= \dfrac{2z - 26}{z^2 - 5z} \cdot \dfrac{x^2 + 3z}{3z - 39}$ Change division to multiplication.

$= \dfrac{2(\cancel{z-13})}{\cancel{z}(z - 5)} \cdot \dfrac{\cancel{z}(z + 3)}{3(\cancel{z-13})},\ z \neq 0,\ 5,\ -3,\ 13$ Factor, restrict the variable, and reduce. When restricting the variable, recall that in $\dfrac{A}{B} \div \dfrac{C}{D}$, $BCD \neq 0$.

$= \dfrac{2(z + 3)}{3(z - 5)} = \dfrac{2z + 6}{3z - 15},\ z \neq 0,\ 5,\ -3,\ 13$ Multiply.

C

Multiply or divide and simplilfy. You need not state the variable restrictions:

33. $\dfrac{5y + 10}{y^2 - 4} \cdot \dfrac{y^2 - 7y + 10}{8y - 40}$

$= \dfrac{5(\cancel{y+2})}{(\cancel{y+2})(y - 2)} \cdot \dfrac{(\cancel{y-5})(\cancel{y-2})}{8(\cancel{y-5})} = \dfrac{5}{8}$ Factor and reduce.

37. $\dfrac{x^2 + 9x + 18}{x^2 + 2x - 15} \div \dfrac{2x + 12}{3x + 15}$

$\quad = \dfrac{x^2 + 9x + 18}{x^2 + 2x - 15} \cdot \dfrac{3x + 15}{2x + 12}$ Invert and multiply.

$\quad = \dfrac{(\cancel{x + 6})(x + 3)}{(\cancel{x + 5})(x - 3)} \cdot \dfrac{3(\cancel{x + 5})}{2(\cancel{x + 6})}$ Factor and reduce.

$\quad = \dfrac{3(x + 3)}{2(x - 3)} = \dfrac{3x + 9}{2x - 6}$ Multiply.

41. $\dfrac{6x^2 + 39x + 63}{10x^2 - 17x + 3} \div \dfrac{6x^2 + 33x + 45}{10x^2 + 33x - 7}$

$\quad = \dfrac{6x^2 + 39x + 63}{10x^2 - 17x + 3} \cdot \dfrac{10x^2 + 33x - 7}{6x^2 + 33x + 45}$ Invert and multiply.

$\quad = \dfrac{(\cancel{3x + 9})(2x + 7)}{(\cancel{5x - 1})(2x - 3)} \cdot \dfrac{(\cancel{5x - 1})(2x + 7)}{(\cancel{3x + 9})(2x + 5)}$ Factor and reduce.

$\quad = \dfrac{(2x + 7)(2x + 7)}{(2x - 3)(2x + 5)} = \dfrac{4x^2 + 28x + 49}{4x^2 + 4x - 15}$ Multiply.

Multiply or divide and simplify:

45. $\dfrac{3y^2 + 6y + 12}{9y - y^3} \cdot \dfrac{y^3 + 9y}{6^3 - 8}$

$\quad = \dfrac{3(\cancel{y^2 + 2y + 4})}{y(3 - y)(3 + y)} \cdot \dfrac{y(y^2 + 9)}{(y - 2)(\cancel{y^2 + 2y + 4})}$ Factor. Divide out common factors.

$\quad = \dfrac{3(y^2 + 9)}{(y - 2)(3 - y)(3 + y)}$

E   Maintain Your Skills

Multiply:

49. $(2xy^2z)^2(3xy^3)^3$

$\quad = [2^2x^2(y^2)^2z^2][3^3x^3(y^3)^3]$

$\quad = 4 \cdot x^2 \cdot y^4 \cdot z^2 \cdot 27 \cdot x^3y^9$

$\quad = (4 \cdot 27)(x^2 \cdot x^3)(y^4 \cdot y^9)(z^2)$

$\quad = 108x^5y^{13}z^2$

53. $(x + 7)(x^3 + 3x - 5)$

$= (x + 7) \cdot x^3 + (x + 7) \cdot 3x + (x + 7) \cdot (-5)$

$= x^4 + 7x^3 + 3x^2 + 21x - 5x - 35$

$= x^4 + 7x^3 + 3x^2 + 16x - 35$

57. $(12x + 13y)^2$

$= (12x)^2 + 2(12x)(13y) + (13y)^2$

$= 144x^2 + 312xy + 169y^2$

EXERCISES 3.3   ADDITION AND SUBTRACTION OF RATIONAL EXPRESSIONS

A

Find the LCD of the fractions:

1. $\dfrac{1}{5ab} + \dfrac{1}{5ac}$

   $5ab = 5 \cdot a \cdot b$ Factor each denominator.
   $5ac = 5 \cdot a \cdot c$ The different factors are 5, a, b, and c.

   LCD $= 5abc$ The LCD is the product of the highest power of each factor.

Add or subtract. Reduce, if possible. You need not state the variable restrictions:

5. $\dfrac{3}{x} + \dfrac{6}{x} - \dfrac{5}{x}$ The fractions have a common denominator.

   $= \dfrac{3 + 6 - 5}{x} = \dfrac{4}{x}$ Combine the numerators and retain the common denominator.

9. $\dfrac{a + 1}{3} - \dfrac{a + 2}{2}$

   LCD $= 6$

   $= \dfrac{(a + 1)(2)}{3(2)} - \dfrac{(a + 2)(3)}{2(3)}$ Build the fractions so that each has the LCD for the denominator.

   $= \dfrac{2a + 2}{6} - \dfrac{3a + 6}{6}$ Multiply.

   $= \dfrac{(2a + 2) - (3a + 6)}{6}$ Subtract the numerators, and retain the common denominator.

   $= \dfrac{2a + 2 - 3a - 6}{6}$

   $= \dfrac{-a - 4}{6}$

13. $\dfrac{1}{5ab} + \dfrac{1}{5ac}$

LCD = 5abc

$\dfrac{1}{5ab} + \dfrac{1}{5ac} = \dfrac{1}{5ab} \cdot \dfrac{c}{c} + \dfrac{1}{5ac} \cdot \dfrac{b}{b}$ Build the fractions so that they have the LCD as the denominator.

$= \dfrac{c}{5abc} + \dfrac{b}{5abc}$ Multiply.

$= \dfrac{b + c}{5abc}$ Add the numerators and retain the common denominator.

17. $\dfrac{5}{2y - 2} + \dfrac{4}{3y - 3}$

LCD = 6(y − 1)   Find the LCD of the fractions.

$\dfrac{5}{2(y - 1)} + \dfrac{4}{3(y - 1)} = \dfrac{5}{2(y - 3)} \cdot \dfrac{3}{3} + \dfrac{4}{3(y - 1)} \cdot \dfrac{2}{2}$ Build the fractions to have the LCD as denominators.

$= \dfrac{15}{6(y - 1)} + \dfrac{8}{6(y - 1)}$ Multiply.

$= \dfrac{23}{6(y - 1)}$

B

Find the LCD of the following fractions:

21. $\dfrac{5x}{x - 2} + \dfrac{6}{2 - x}$

LCD = x − 2 or 2 − x   The denominators are opposites, so either can be the LCD.

Add or subtract. Simplify, if possible. You need not state the variable restrictions:

25. $\dfrac{7}{w - 4} + \dfrac{1}{4 - w}$

LCD = w − 4

$\dfrac{7}{w - 4} + \dfrac{1}{4 - w} = \dfrac{7}{w - 4} + \dfrac{1}{4 - w} \cdot \dfrac{-1}{-1}$ Build each fraction using the LCD as denominator.

$= \dfrac{7}{w - 4} + \dfrac{-1}{w - 4}$ Multiply.

$= \dfrac{6}{w - 4}$ Add.

or
LCD = 4 - w

$$\frac{7}{w-4} + \frac{1}{4-w} = \frac{7}{w-4} \cdot \frac{-1}{-1} + \frac{1}{4-w}$$
$$= \frac{-7}{4-w} + \frac{1}{4-w} = \frac{-6}{4-w}$$

29. $\frac{3t+4}{5} + t - 1$

LCD = 5

$$\frac{3t+4}{5} + t - 1 = \frac{3t+4}{5} + t \cdot \frac{5}{5} - 1 \cdot \frac{5}{5} \quad \text{Use the LCD to build each term.}$$

$$= \frac{3t+4}{5} + \frac{5t}{5} - \frac{5}{5} \quad \text{Multiply.}$$

$$= \frac{3t+4+5t-5}{5}$$

$$= \frac{8t-1}{5}$$

33. $\frac{3}{x+6} + \frac{5}{2x-3}$

LCD = (x + 6)(2x - 3)

$$\frac{3}{x+6} + \frac{5}{2x-3} = \frac{3}{x+6} \cdot \frac{2x-3}{2x-3} + \frac{5}{2x-3} \cdot \frac{x+6}{x+6} \quad \text{Build using the LCD as the denominator.}$$

$$= \frac{6x-9}{(x+6)(2x-3)} + \frac{5x+30}{(x+6)(2x-3)} \quad \text{Multiply.}$$

$$= \frac{11x+21}{(x+6)(2x-3)} \quad \text{Add.}$$

Add or subtract. Reduce, if possible. State the restrictions on the variables:

37. $\frac{1}{2x-3} - \frac{x+6}{x+3}$

$x \neq \frac{3}{2}, -3;$ LCD = (2x - 3)(x + 3)   Restrict the variable.

$$\frac{1}{2x-3} - \frac{x+6}{x+3} = \frac{1}{2x-3} \cdot \frac{x+3}{x+3} - \frac{x+6}{x+3} \cdot \frac{2x-3}{2x-3}$$

$$= \frac{x+3}{(2x-3)(x+3)} + \frac{-1(2x^2+9x-18)}{(x+3)(2x-3)}$$

$$= \frac{-2x^2-8x+21}{(2x-3)(x+3)}, \quad x \neq \frac{3}{2}, -3$$

C

Find the LCD of the following fractions:

41. $\dfrac{x+1}{x^2+2x-15} + \dfrac{2x+1}{x^2-4x+3}$

$x^2 + 2x - 15 = (x+5)(x-3)$ Factor each denominator.
$x^2 - 4x + 3 = (x-3)(x-1)$
LCD $= (x-1)(x-3)(x+5)$ The LCD is the product of the highest power of each factor.

Add or subtract. Reduce, if possible. State the restrictions on the variables:

45. $\dfrac{2}{x+3} - \dfrac{3}{x-5} + \dfrac{x}{x^2-2x-15}$

$= \dfrac{2}{x+3} - \dfrac{3}{x-5} + \dfrac{x}{(x-5)(x+3)}$ Factor the third denominator.

$x \neq -3, 5$; LCD $= (x+3)(x-5)$ Restrict the variable.

$\dfrac{2}{x+3} - \dfrac{3}{x-5} + \dfrac{x}{(x+3)(x-5)}$

$= \dfrac{2}{x+3} \cdot \dfrac{x-5}{x-5} - \dfrac{3}{x-5} \cdot \dfrac{x+3}{x+3} + \dfrac{x}{(x+3)(x-5)}$

$= \dfrac{2(x-5)}{(x+3)(x-5)} + \dfrac{-3(x+3)}{(x+3)(x-5)} + \dfrac{x}{(x+3)(x-5)}$

$= \dfrac{2x-10}{(x+3)(x-5)} + \dfrac{-3x-9}{(x+3)(x-5)} + \dfrac{x}{(x+3)(x-5)}$

$= \dfrac{-19}{(x+3)(x-5)}, \; x \neq -3, 5$

49. $\dfrac{6}{x^2 + x - 6} - \dfrac{5}{x^2 + 2x - 8} + \dfrac{4}{x^2 + 7x + 12}$

$\quad\quad x^2 + x - 6 = (x + 3)(x - 2)$ Factor each denominator.
$\quad\quad x^2 + 2x - 8 = (x + 4)(x - 2)$
$\quad\quad x^2 + 7x + 12 = (x + 4)(x + 3)$
$\quad\quad x \neq -3, 2, -4$ Restrict the variable.

$\quad\quad$ LCD $= (x + 3)(x - 2)(x + 4)$

$\dfrac{6}{(x + 3)(x - 2)} \cdot \dfrac{x + 4}{x + 4} - \dfrac{-5}{(x + 4)(x - 2)} \cdot \dfrac{x + 3}{x + 3} + \dfrac{4}{(x + 4)(x + 3)} \cdot \dfrac{x - 2}{x - 2}$

$= \dfrac{6(x + 4)}{(x + 3)(x - 2)(x + 4)} + \dfrac{-5(x + 3)}{(x + 4)(x - 2)(x + 3)}$
$\quad + \dfrac{4(x - 2)}{(x + 4)(x + 3)(x - 2)}$

$= \dfrac{6x + 24}{(x + 3)(x - 2)(x + 4)} + \dfrac{-5x - 15}{(x + 3)(x - 2)(x + 4)}$
$\quad + \dfrac{4x - 8}{(x + 3)(x - 2)(x + 4)}$

$= \dfrac{5x + 1}{(x + 3)(x - 2)(x + 4)},\ x \neq -3, 2, -4$

53. $\dfrac{1}{x} - \dfrac{2}{x + 2} - \dfrac{2}{x^2 + 3x + 2}$

$\quad x = 0, -2, -1;\ \text{LCM} = x(x + 2)(x + 1)$

$\dfrac{1(x + 2)(x + 1)}{x(x + 2)(x + 1)} - \dfrac{2(x)(x + 1)}{(x + 2)(x)(x + 1)} - \dfrac{2(x)}{(x + 2)(x + 1)(x)}$

$= \dfrac{x^2 + 3x + 2}{x(x + 2)(x + 1)} + \dfrac{-(2x^2 + 2x)}{x(x + 2)(x + 1)} + \dfrac{-(2x)}{x(x + 2)(x + 1)}$

$= \dfrac{x^2 + 3x + 2 - 2x^2 - 2x - 2x}{x(x + 1)(x + 2)}$

$= \dfrac{-x^2 - x + 2}{x(x + 1)(x + 2)} = \dfrac{-1(x^2 + x - 2)}{x(x + 1)(x + 2)}$ Factor and reduce.

$= \dfrac{-1(x - 1)\cancel{(x + 2)}}{x\cancel{(x + 2)}(x + 1)}$

$= \dfrac{-x + 1}{x(x + 1)},\ x = 0, -1$

57. $\dfrac{p}{q^{-1}} - \dfrac{p^{-1}}{q}$

$\dfrac{pq}{1} - \dfrac{1}{pq}$ Rewrite in fraction form.
Recall that $x^{-m} = \dfrac{1}{x^m}$.

$p \neq 0,\ q \neq 0;\ \text{LCM} = pq$

$\dfrac{(pq)(pq)}{1(pq)} - \dfrac{1}{pq}$ Build each fraction.

$= \dfrac{p^2q^2}{pq} - \dfrac{1}{pq} = \dfrac{p^2q^2 - 1}{pq},\ p \neq 0,\ q \neq 0$

E Maintain Your Skills

Factor:

61. $a^2 - 13a + 42$

  $mn = 42 \qquad m + n = -13$
  $-1(-42) \qquad\qquad -43$
  $-2(-21) \qquad\qquad -23$
  $-3(-14) \qquad\qquad -17$
  $-6(-7) \qquad\qquad\ -13$

  Find two numbers whose product is 42 and whose sum is -13.

  $(a - 6)(a - 7)$

65. $4m^2n^2 - 49p^2$

  $(2mn)^2 - (7p)^2$ Difference of squares.
  $(2mn - 7p)(2mn + 7p)$ Conjugate pairs.

Solve:

69. $10x^2 - 7x - 45 = 0$
  $(5x + 9)(2x - 5) = 0$ Factor.
  $5x + 9 = 0 \text{ or } 2x - 5 = 5$ Zero-product law.
  $x = -\dfrac{9}{5} \text{ or } \quad x = \dfrac{5}{2}$

  The solution set is $\left\{-\dfrac{9}{5},\ \dfrac{5}{2}\right\}$.

EXERCISES 3.4   COMPLEX FRACTIONS

Simplify. You need not state variable restrictions:

A

1. $\dfrac{\frac{3}{x}}{\frac{6}{y}} = \dfrac{3}{x} \div \dfrac{6}{y}$ Rewrite as a division problem.

$= \dfrac{3}{x} \cdot \dfrac{y}{6} = \dfrac{y}{2x}$ Multiply by the reciprocal of the divisor and simplify.

5. $\dfrac{\frac{1}{x+1}}{\frac{1}{x-1}} = \dfrac{(x+1)(x-1)\left[\frac{1}{x+1}\right]}{(x+1)(x-1)\left[\frac{1}{x-1}\right]}$ Multiply the numerator and denominator by the LCM of $(x+1)$ and $(x-1)$ which is $(x+1)(x-1)$.

$= \dfrac{x-1}{x+1}$ Simplify.

9. $\dfrac{q - \frac{1}{3}}{p + \frac{1}{3}} = \dfrac{3\left[q - \frac{1}{3}\right]}{3\left[p + \frac{1}{3}\right]}$ Multiply by the LCM, 3.

$= \dfrac{3q - 1}{3p + 1}$ Simplify.

Simplify. State the restrictions on the variables:

13. $\dfrac{2 - \frac{1}{a}}{3 + \frac{1}{a}} = \dfrac{a\left[2 - \frac{1}{a}\right]}{a\left[3 + \frac{1}{a}\right]}, \; a \ne 0$ Multiply by the LCM, a.

$= \dfrac{2a - 1}{3a + 1}, \; a \ne 0, \; -\dfrac{1}{3}$ Simplify.

B

Simplify. You need not state variable restrictions:

17. $\dfrac{\frac{x}{3} + 2}{\frac{x+1}{3}} = \left(\dfrac{x}{3} + 2\right) \div \dfrac{x+1}{3}$ Rewrite as a division problem.

$= \dfrac{x+6}{3} \div \dfrac{x+1}{3}$ Express the dividend, $\dfrac{x}{3} + 2$, using 3 as the LCD.

$= \dfrac{x+6}{3} \cdot \dfrac{3}{x+1}$ Multiply by the reciprocal of the divisor.

$= \dfrac{x+6}{x+1}$ Simplify.

21. $\dfrac{t - \frac{t^2}{w}}{t + \frac{t}{w^2}} = \dfrac{w^2\left(t - \frac{t^2}{w}\right)}{w^2\left(t + \frac{t}{w^2}\right)}$ Multiply by the LCM, w^2.

$= \dfrac{tw^2 - t^2w}{tw^2 + t}$ Simplify.

$= \dfrac{tw(w - t)}{t(w^2 + 1)}$ Factor both numerator and denominator to identify any possible common factors. Divide out the common factor, t.

$= \dfrac{w(w - t)}{w^2 + 1}$

$= \dfrac{w^2 - tw}{w^2 + 1}$ Multiply.

25. $\dfrac{\dfrac{5y-8}{3}}{\dfrac{y+1}{4} - \dfrac{y-1}{6}}$

$= \dfrac{5y-8}{3} \div \left[\dfrac{y+1}{4} - \dfrac{y-1}{6}\right]$ Rewrite the divisor using 12 as the LCD.

$= \dfrac{5y-8}{3} \div \left[\dfrac{y+1}{4}\cdot\dfrac{3}{3} - \dfrac{y-1}{6}\cdot\dfrac{2}{2}\right]$

$= \dfrac{5y-8}{3} \div \left[\dfrac{3y+3}{12} + \dfrac{-2y+2}{12}\right]$

$= \dfrac{5y-8}{3} \div \dfrac{y+5}{12}$ Simplify the divisor by adding.

$= \dfrac{5y-8}{\underset{1}{\cancel{3}}} \cdot \dfrac{\overset{4}{\cancel{12}}}{y+5}$ Multiply by the reciprocal of the divisor.

$= \dfrac{20y-32}{y+5}$ Simplify.

Simplify. State the restrictions on the variables:

29. $\dfrac{2a^{-1} - 3b^{-1}}{b - a}$

$= \dfrac{\dfrac{2}{a} - \dfrac{3}{b}}{b - a}$ Rewrite the negative exponents.

$= \dfrac{ab\left[\dfrac{2}{a} - \dfrac{3}{b}\right]}{ab(b - a)}$ Multiply by the LCD, ab.

$= \dfrac{2b - 3a}{ab(b - a)},\ a \neq 0,\ b \neq 0,\ a \neq b$ Multiply. Restrict the variables.

C

Simplify. You need not state the variable restrictions:

33. $\dfrac{\dfrac{1}{a+b} - \dfrac{1}{a-b}}{\dfrac{1}{a+b} + \dfrac{1}{a-b}} = \dfrac{(a+b)(a-b)\left[\dfrac{1}{a+b} - \dfrac{1}{a-b}\right]}{(a+b)(a-b)\left[\dfrac{1}{a+b} + \dfrac{1}{a-b}\right]}$ Muliply by the LCD, $(a+b)(a-b)$.

$= \dfrac{(a-b) - (a+b)}{(a-b) + (a+b)}$ Simplify.

$= \dfrac{a-b-a-b}{a-b+a+b} = \dfrac{-2b}{2a} = -\dfrac{b}{a}$

37. $$\dfrac{\dfrac{2}{a-1} + \dfrac{1}{a+1}}{\dfrac{6}{a^2-1}}$$

$$\dfrac{(a+1)(a-1)\left[\dfrac{2}{a-1} + \dfrac{1}{a+1}\right]}{(a+1)(a-1)\left[\dfrac{6}{a^2-1}\right]}$$

Multiply both numerator and denominator by the LCD which is $(a+1)(a-1)$.

$$= \dfrac{2(a+1) + 1(a-1)}{6}$$

Simplify.

$$= \dfrac{2a+2+a-1}{6} = \dfrac{3a+1}{6}$$

41. $((x+2)^{-1} - (x-2)^{-1})^{-1}$

$$\left[\dfrac{1}{x+2} - \dfrac{1}{x-2}\right]^{-1}$$

Rewrite the negative exponents of the inner expressions using fractions.

$$= \left[\dfrac{1}{x+2} \cdot \dfrac{x-2}{x-2} - \dfrac{1}{x-2} \cdot \dfrac{x+2}{x+2}\right]^{-1}$$

Build each fraction to have the LCD, $(x+2)(x-2)$, as its denominator.

$$= \left[\dfrac{x-2}{(x+2)(x-2)} - \dfrac{x+2}{(x-2)(x+2)}\right]^{-1}$$

$$= \left[\dfrac{x-2-x-2}{(x+2)(x-2)}\right]^{-1} = \left[\dfrac{-4}{(x+2)(x-2)}\right]^{-1}$$

$$= \dfrac{(x+2)(x-2)}{-4}$$

Rewrite the negative exponent as a reciprocal.

$$= \dfrac{x^2-4}{-4}$$

Simplify.

or $\dfrac{-1(x^2-4)}{-1(-4)} = \dfrac{4-x^2}{4}$

45. $\dfrac{2a + \dfrac{1}{6 - \dfrac{1}{a}}}{3a + \dfrac{2}{6 + \dfrac{1}{a}}}$

$\dfrac{2a + \dfrac{1}{6 - \dfrac{1}{a}} \cdot \dfrac{a}{a}}{3a + \dfrac{2}{6 + \dfrac{1}{a}} \cdot \dfrac{a}{a}}$ To eliminate the complex fraction in the numerator and denominator, multiply by $\dfrac{a}{a}$.

$= \dfrac{2a + \dfrac{a}{6a - 1}}{3a + \dfrac{2a}{6a + 1}}$ Simplify.

$= \dfrac{(6a - 1)(6a + 1)\left[2a + \dfrac{1}{6a - 1}\right]}{(6a - 1)(6a + 1)\left[3a + \dfrac{2a}{6a + 1}\right]}$ Multiply the numerator and denominator by $(6a - 1)(6a + 1)$, the LCD.

$= \dfrac{2a(36a^2 - 1) + a(6a + 1)}{3a(36a^2 - 1) + 2a(6a - 1)}$ Distributive law.

$= \dfrac{72a^3 - 2a + 6a^2 + a}{108a^3 - 3a + 12a^2 - 2a}$ Multiply.

$= \dfrac{72a^3 + 6a^2 - a}{108a^3 + 12a^2 - 5a}$ Combine like terms.

$= \dfrac{a(72a^2 + 6a - 1)}{a(108a^2 + 12a - 5)}$ Factor out the common monomial, a.

$= \dfrac{\cancel{a}(12a - 1)(6a + 1)}{\cancel{a}(6a - 1)(18a + 5)}$

$= \dfrac{(12a - 1)(6a + 1)}{(6a - 1)(18a + 5)}$

E Maintain Your Skills

Factor:

49. $81y^2 - 198y + 121$ Perfect square trinomial.
    $= (9y - 11)^2$

53. $64x^3 - 1$ Difference of cubes.
    $= (4x - 1)(16x^2 + 4x + 1)$

Solve:

57. $2x^2 - 35x + 17 = 0$
    $(2x - 1)(x - 17) = 0$ Factor.
    $2x - 1 = 0$ or $x - 17 = 0$ Zero-product law.
        $x = \frac{1}{2}$ $x = 17$

The solution set is $\left\{\frac{1}{2}, 17\right\}$.

EXERCISES 3.5   EQUATIONS CONTAINING RATIONAL EXPRESSIONS

A

Solve:

1. $\frac{x}{4} + \frac{1}{3} = \frac{x}{12}$ There are no restrictions on the variable.

   $12\left[\frac{x}{4} + \frac{1}{3}\right] = 12\left[\frac{x}{12}\right]$ Multiply each side of the equation by the LCD of 4, 3, and 12. LCD = 12.

   $3x + 4 = x$ Simplify.
   $2x = -4$
   $x = -2$

   The solution set is $\{-2\}$.

5. $\frac{5}{t - 2} = 4$

   $(t - 2)\left[\frac{5}{t - 2}\right] = (t - 2)4$ Multiply each side of the equation by the LCD, $t - 2$.
   $5 = 4t - 8$ Simplify.
   $13 = 4t$
   $\frac{13}{4} = t$

   The solution set is $\left\{\frac{13}{4}\right\}$.

9. $\frac{4}{x + 3} = \frac{5}{2x + 7}$

   $(x + 3)(2x + 7)\left[\frac{4}{x + 3}\right] = (x + 3)(2x + 7)\left[\frac{5}{2x + 7}\right]$ Multiply each side by the LCD: $(x+3)(2x+7)$.

   $4(2x + 7) = 5(x + 3)$
   $8x + 28 = 5x + 15$
   $3x = -13$
   $x = -\frac{13}{3}$

   The solution set is $\left\{-\frac{13}{3}\right\}$.

13.
$$\frac{4}{x-3} + \frac{3}{x-3} = 5$$

$$x - 3\left[\frac{4}{x-3} + \frac{3}{x-3}\right] = (x-3)5$$
$$4 + 3 = 5x - 15$$
$$22 = 5x$$
$$\frac{22}{5} = x$$

The solution set is $\left\{\frac{22}{5}\right\}$.

B

17.
$$\frac{1}{x+2} - \frac{3}{4} = \frac{7}{x+2}$$

$$4(x+2)\left[\frac{1}{x+2} - \frac{3}{4}\right] = 4(x+2)\left[\frac{7}{x+2}\right] \quad \text{Multiply each side by the LCD of 4 and } x+2.$$
$$\text{LCD} = 4(x+2).$$
$$4 - 3(x+2) = 4(7)$$
$$4 - 3x - 6 = 28$$
$$-3x = 30$$
$$x = 10$$

The solution set is {-10}.

21.
$$\frac{3}{y+3} - \frac{9}{y^2-9} = \frac{2}{y-3}$$

$$(y+3)(y-3)\left[\frac{3}{y+3} - \frac{9}{y^2-9}\right] = (y+3)(y-3)\left[\frac{2}{y-3}\right]$$

Multiply each side by the LCM of $y+3$, y^2-9, and $y-3$. LCM = $(y+3)(y-3)$ or y^2-9.

$$3(y-3) - 9 = 2(y+3)$$
$$3y - 9 - 9 = 2y + 6$$
$$y = 24$$

The solution set is {24}.

25. $$\frac{4}{x-3} - \frac{8}{x^2-9} = \frac{3}{3-x}$$

$$\frac{4}{x-3} - \frac{8}{x^2-9} = \frac{-1(3)}{-1(3-x)}$$ Multiply the numerator and denominator by -1.

$$\frac{4}{x-3} - \frac{8}{x^2-9} = \frac{-3}{x-3}$$

$$(x+3)(x-3)\left[\frac{4}{x-3} - \frac{8}{x^2-9}\right] = (x+3)(x-3)\left[\frac{-3}{x-3}\right]$$

Now multiply both sides by the LCM: $x^2 - 9$ or $(x+3)(x-3)$. Simplify.

$$4(x-3) - 8 = -3(x+3)$$
$$4x + 12 - 8 = -3x - 9$$
$$7x = -13$$
$$x = -\frac{13}{7}$$

The solution set is $\left\{-\frac{13}{7}\right\}$.

29. Solve for a: $\dfrac{3}{b+a} - \dfrac{2}{b} = 5$

$$b(b+1)\left[\frac{3}{b+a} - \frac{2}{b}\right] = b(b+a)(5)$$ Multiply both sides by the LCM, $b(b+a)$.
$$3b - 2(b+a) = 5(b^2 + ab)$$
$$3b - 2b - 2a = 5b^2 + 5ab$$
$$-2a - 5ab = 5b^2 - b$$
$$a(-2 - 5b) = 5b^2 - b$$ On the left, factor out an a.
$$a = \frac{5b^2 - b}{-2 - 5b} \cdot \frac{-1}{-1}$$
$$a = \frac{b - 5b^2}{5b + 2}$$

C

33. $$\frac{1}{x^2+4x+3} - \frac{1}{x^2-5x-6} = \frac{1}{x^2-3x-18}$$

$$\frac{1}{(x+3)(x+1)} - \frac{1}{(x-6)(x+1)} = \frac{1}{(x-6)(x+3)}$$

$$(x-6) - (x+3) = x+1$$ Multiply both sides
$$x - 6 - x - 3 = x+1$$ by the LCD,
$$-9 = x+1$$ $(x+3)(x+1)(x-6)$, and
$$-10 = x$$ simplify.

The solution set is $\{-10\}$.

37. $$\frac{x+3}{x^2-x-2} + \frac{x+6}{x^2+3x+2} = \frac{2x-1}{x^2-4}$$

$\frac{x+3}{(x-2)(x+1)} + \frac{x+6}{(x+2)(x+1)} = \frac{2x-1}{(x-2)(x+2)}$ The LCD is $(x-2)(x+1)(x+2)$.

$(x+2)(x+3) + (x+6)(x-2) = (2x-1)(x+1)$ Multiply by the
$x^2 + 5x + 6 + x^2 + 4x - 12 = 2x^2 + x - 1$ LCD and
$9x - 6 = x - 1$ simplify.
$8x - 6 = -1$
$8x = 5$
$x = \frac{5}{8}$

The solution set is $\left\{\frac{5}{8}\right\}$.

41. Frank and Freda can pick a quart of blueberries in 12 minutes. Frank alone takes 18 minutes longer than Freda. How long does it take each of them to pick a quart?

Simpler word form:
$\begin{bmatrix}\text{Fraction of work}\\ \text{done by Frank}\\ \text{in one minute}\end{bmatrix} + \begin{bmatrix}\text{Fraction of work}\\ \text{done by Freda}\\ \text{in one minute}\end{bmatrix} = \begin{bmatrix}\text{Fraction of work}\\ \text{done by both}\\ \text{in one minute}\end{bmatrix}$

Select variable:
Let t represent the time it takes Freda to pick a quart of berries.

	Time alone	Fraction done in one minute
Freda	t	$\frac{1}{t}$
Frank	t + 18	$\frac{1}{t+18}$
Both	12	$\frac{1}{12}$

Translate to algebra:
$\frac{1}{t+18} + \frac{1}{t} = \frac{1}{12}$
$12t + 12(t+18) = t(t+18)$ Multiply each side by the
$12t + 12t + 216 = t^2 + 18t$ LCD: $12t(t+18)$.
$t^2 - 6t - 216 = 0$
$(t-18)(t+12) = 0$
$t - 18 = 0$ or $t + 12 = 0$
$t = 18$ or $t = -12$ Reject $t = -12$ as t represents a positive number.

Frank takes 18 minutes longer than Freda, therefore, 18 (Freda's time) + 18 = 36.

It takes Freda 18 minutes to pick a quart of blueberries, and it takes Frank 36 minutes.

45. Charlie and Jim decide to drive to Seattle. They both leave from the same place at the same time, but Charlie drives 5 miles/hr faster than Jim. If Charlie has traveled 65 miles in the same time that Jim has traveled 60 miles, find how fast each was driving.

Simpler word form:
$$\begin{bmatrix} \text{Time it takes Charlie} \\ \text{to travel 65 miles} \end{bmatrix} = \begin{bmatrix} \text{Time it takes Jim} \\ \text{to travel 60 miles} \end{bmatrix}$$

Select a variable:
Let r represent the rate of speed that Jim travels. Then r + 5 represents the rate of speed that Charlie travels.

	Distance traveled	Rate of Speed	Time
Jim	60	r	$\frac{60}{r}$
Charlie	65	r + 5	$\frac{65}{r+5}$

Translate to algebra:

$\frac{60}{r} = \frac{65}{r+5}$

60(r + 5) = 65r	Cross multiply.
60r + 300 = 65r	Simplify.
300 = 5r	
r = 60	Jim's rate of speed.

Charlie travels 5 miles faster
than Jim so 60 + 5 = 65.   Charlie's rate of speed.

Charlie was traveling at 65 mph, and Jim was traveling at 60 mph.

49. Ariel saved $900 for vacation. She figures that by spending $15 a day less, she can stay away 3 days longer than originally planned. How many days did she originally plan to be gone?

(Hint: $\frac{\text{total cost}}{\text{number of days}}$ = cost per day)

Simpler word form:
$$\begin{bmatrix} \frac{\text{total cost}}{\text{number of days}} \\ \text{originally planned} \end{bmatrix} - 15 = \begin{bmatrix} \frac{\text{total cost}}{\text{number of days}} \\ \text{in new plan} \end{bmatrix}$$

Select a variable:
Let x represent the number of days Ariel originally planned to be gone. Then x + 3 represents the number of days she will be away in her new plan.

Translate to algebra:
$$\frac{900}{x} - 15 = \frac{900}{x + 3}$$

Solve:

$x(x + 3)\left[\dfrac{900}{x} - 15\right] = x(x + 3)\left[\dfrac{900}{x + 3}\right]$ Multiply each side by the LCM: $x(x + 3)$.

$900(x + 3) - 15(x)(x + 3) = 900x$ Simplify.
$900 + 2700 - 15x^2 - 45x = 900x$
$-15x^2 - 45x + 2700 = 0$
$15x^2 + 45x - 2700 = 0$ Multiply each side by -1.
$15(x^2 + 3x - 180) = 0$ Factor.
$(x + 15)(x - 12) = 0$ Divide each side by 15 and factor.

$x + 15 = 0$ or $x - 12 = 0$ Zero-product law.
$x = -15$ $x = 12$ Reject -15 because it cannot represent the number of vacation days.

So Ariel orginally planned to be away for 12 days.

E Maintain Your Skills

Solve for w:

53. $8w - 1 \leq w + 16$
     $7w \leq 17$
     $w \leq \dfrac{17}{7}$

In interval notation, the solution set is $\left(-\infty, \dfrac{17}{7}\right]$.

Multiply:

57. $-8xyz(6x^2 - 3y + 2yz - 2z^2)$
     $= (-8xyz)(6x^2) + (-8xyz)(-3y) + (-8xyz)(2yz) + (-8xyz)(-2z^2)$
     $= -48x^3yz + 24xy^2z - 16xy^2z^2 + 16xyz^3$

EXERCISES 3.6   DIVISION OF POLYNOMIALS

A

Divide:

1.
$$\begin{array}{r} a \\ a^2 - 3 \overline{) a^2 - 3a} \\ \underline{a^2 - 3a} \\ 0 \end{array}$$

a times a - 3 equals $a^2 - 3a$. Place the a in the answer above -3a and multiply. Subtract. The remainder is 0.

    The quotient is a.

5.
$$\begin{array}{r} 3x \\ 6x + 11 \overline{) 18x^2 + 33x} \\ \underline{18x^2 + 33x} \\ 0 \end{array}$$

3x times 6x - 11 equals $18x^2 - 33x$. Place the 3x in the answer above 33x. Multiply. Subtract.

    The quotient is 3x.

9.
$$\begin{array}{r} x + 3 \\ x + 2 \overline{) x^2 + 5x + 6} \\ \underline{x^2 + 2x} \\ 3x + 6 \\ \underline{3x + 6} \\ 0 \end{array}$$

x times x equals x^2. Place the x in the answer above 5x. Multiply. Subtract.
3 times x equals 3x. Place the 3 in the answer above 6, and multiply. Subtract.

    The quotient is x + 3.

13. $(x^2 + 7x - 18) \div (x + 9)$

$$\begin{array}{r} x - 2 \\ x + 9 \overline{) x^2 + 7x - 18} \\ \underline{x^2 + 9x} \\ -2x - 18 \\ \underline{-2x - 18} \\ 0 \end{array}$$

Rewrite as a long division problem. $x \cdot x = x^2$. $-2 \cdot x = -2x$.

    The quotient is x - 2.

B

17.
$$\begin{array}{r} 2x + 7 \\ x - 3 \overline{) 2x^2 + x - 21} \\ \underline{2x^2 - 6x} \\ 7x - 21 \\ \underline{7x - 21} \\ 0 \end{array}$$

$2x \cdot x = 2x^2$.

$7 \cdot x = 7x$.

    The quotient is 2x + 7

97

21.
$$\begin{array}{r} 3x + 7 \\ 2x + 3 \overline{)6x^2 + 23x + 21} \\ \underline{6x^2 + 9x } \\ 14x + 21 \\ \underline{14x + 21} \\ 0 \end{array}$$
3x times 2x equals $6x^2$.

7 times 2x equals 14x.

The qotient is $3x + 7$.

25. $(2x^2 - 5x + 16) \div (2x + 1)$

$$\begin{array}{r} x - 3 \\ 2x + 1 \overline{)2x^2 - 5x + 16} \\ \underline{2x^2 + x } \\ -6x + 16 \\ \underline{-6x - 3} \\ 19 \end{array}$$

The remainder is 19.

So $(2x^2 - 5x + 16) \div (2x + 1) = x - 3 + \dfrac{19}{2x + 1}$.

29. $(x^3 + 3x^2 + 3x + 1) \div (x + 1)$

$$\begin{array}{r} x^2 + 2x + 1 \\ x + 1 \overline{)x^3 + 3x^2 + 3x + 1} \\ \underline{x^3 + x^2 } \\ 2x^2 + 3x + 1 \\ \underline{2x^2 + 2x } \\ x + 1 \\ \underline{x + 1} \\ 0 \end{array}$$

The quotient is $x^2 + 2x + 1$.

C

33.
$$\begin{array}{r} x^2 + 3 \\ x - 2 \overline{)x^3 - 2x^2 + 3x - 6} \\ \underline{x^3 - 2x^2 } \\ 3x - 6 \\ \underline{3x - 6} \\ 0 \end{array}$$

The quotient is $x^2 + 3$.

37.
$$\begin{array}{r} 2x^2 + x - 1 \\ 3x + 1 \overline{\smash{\big)}\, 6x^3 + 5x^2 - 2x + 1} \\ \underline{6x^3 + 2x^2 } \\ 3x^2 - 2x + 1 \\ \underline{3x^2 + x } \\ -3x + 1 \\ \underline{-3x - 1} \\ 2 \end{array}$$

The quotient is $2x^2 + x - 1 + \dfrac{2}{3x + 1}$.

41.
$$\begin{array}{r} x^5 - x^4 + x^3 - x^2 + x - 1 \\ x + 1 \overline{\smash{\big)}\, x^6 + 0x^5 + 0x^4 + 0x^3 + 0x^2 + 0x + 1} \\ \underline{x^6 + x^5 } \\ -x^5 + 0x^4 + 0x^3 + 0x^2 + 0x + 1 \\ \underline{-x^5 - x^4 } \\ x^4 + 0x^3 + 0x^2 + 0x + 1 \\ \underline{x^4 + x^3 } \\ -x^3 + 0x^2 + 0x + 1 \\ \underline{-x^3 - x^2 } \\ x^2 + 0x + 1 \\ \underline{x^2 + x } \\ -x + 1 \\ \underline{-x - 1} \\ 2 \end{array}$$

Insert zeros as place holders for the terms that are missing.

The quotient is $x^5 - x^4 + x^3 - x^2 + x - 1 + \dfrac{2}{x + 1}$.

D

45. Is $x + 5$ a factor of $x^3 + 10x^2 + 36x + 55$? If so, write the factor.

$$\begin{array}{r} x^2 + 5x + 11 \\ x + 5 \overline{\smash{\big)}\, x^3 + 10x^2 + 36x + 55} \\ \underline{x^3 + 5x^2 } \\ 5x^2 + 36x + 55 \\ \underline{5x^2 + 25x } \\ 11x + 55 \\ \underline{11x + 55} \\ 0 \end{array}$$

Yes, $x + 5$ is a factor of $x^3 + 10x^2 + 36x + 55$.
The factors are $x + 5$ and $x^2 + 5x + 11$ or $(x + 5)(x^2 + 5x + 11)$.

E   Maintain Your Skills

Multiply:

49. $(x + 3)(x^2 - 3x + 9)$
$= x(x^2) + x(-3x) + x(9) + 3(x^2) + 3(-3x) + 3(9)$
$= x^3 - 3x^2 + 9x + 3x^2 - 9x + 27$
$= x^3 + 27$

53. $(x^2 + 2x - 1)(x^2 - 2x - 1)$

$$\begin{array}{r} x^2 + 2x - 1 \\ x^2 - 2x - 1 \\ \hline -x^2 - 2x + 1 \\ -2x^3 - 4x^2 + 2x \\ x^4 + 2x^3 - x^2 \\ \hline x^4 \qquad - 6x^2 \qquad + 1 \end{array}$$

Arrange vertically and multiply.

The product is $x^4 - 6x^2 + 1$.

57. Write in scientific notation:

$0.0000502 = 5.02 \times 10^{-5}$

EXERCISES 3.7   SYNTHETIC DIVISION

A

Divide. Use synthetic division.

1. $(x^2 - 2x - 15) \div x - 5$

$$\begin{array}{r|rrr} 5 & 1 & -2 & -15 \\ & & +5 & +15 \\ \hline & 1 & +3 & 0 \end{array}$$

Divide by 5, the opposite of -5.

The quotient is $x + 3$.

5. $(4x^2 - 31x + 55) \div x - 5$

$$\begin{array}{r|rrr} 5 & 4 & -32 & 55 \\ & & 20 & -55 \\ \hline & 4 & -11 & 0 \end{array}$$

Divide by 5, the opposite of -5.

The quotient is $4x - 11$.

9. $x^3 + 2x^2 - 8x + 5) \div x - 1$

$$\begin{array}{r|rrrr} 1 & 1 & 2 & -8 & 5 \\ & & 1 & 3 & -5 \\ \hline & 1 & 3 & -5 & 0 \end{array}$$

Divide by 1.

The quotient is $x^2 + 3x - 5$.

100

B

Divide. Use synthetic division.

13. $(x^2 - 3x + 7) \div (x + 2)$

$$\begin{array}{r|rrr} -2 & 1 & -3 & +7 \\ & & -2 & +10 \\ \hline & 1 & -5 & +17 \end{array}$$

The remainder is 17.

The quotient is $x - 5 + \dfrac{17}{x + 2}$.

17. $(x^3 - 5x^2 + 12x - 15) \div (x - 3)$

$$\begin{array}{r|rrrr} 3 & 1 & -5 & 12 & -15 \\ & & 3 & -6 & 18 \\ \hline & 1 & -2 & 6 & 3 \end{array}$$

The quotient is $x^2 - 2x + 6 + \dfrac{3}{x - 3}$.

21. $(2x^3 - 5x^2 + 4x + 3) \div (x - 1)$

$$\begin{array}{r|rrrr} 1 & 2 & -5 & 4 & 3 \\ & & 2 & -3 & 1 \\ \hline & 2 & -3 & 1 & 4 \end{array}$$

The quotient is $2x^2 - 3x + 1 + \dfrac{4}{x - 1}$.

C

Divide. Use synthetic division.

25. $(x^3 - 3x - 7) \div (x + 2)$

$$\begin{array}{r|rrrr} -2 & 1 & +0 & -3 & -7 \\ & & -2 & +4 & -2 \\ \hline & 1 & -2 & +1 & -9 \end{array}$$

Insert zero for the missing x term.

The missing quotient is $x^2 - 2x + 1 - \dfrac{9}{x + 2}$.

29. $(2x^4 - 13x^3 + 17x^2 + 18x - 24) \div (x - 4)$

$$\begin{array}{r|rrrrr} 4 & 2 & -13 & +17 & +18 & -24 \\ & & +8 & -20 & -12 & +24 \\ \hline & 2 & -5 & -3 & +6 & +0 \end{array}$$

The quotient is $2x^3 - 5x^2 - 3x + 6$

33. $(x^4 - 16) \div (x + 2)$

$$\begin{array}{r|rrrrr} -2 & 1 & +0 & +0 & +0 & -16 \\ & & -2 & +4 & -8 & +16 \\ \hline & 1 & -2 & +4 & -8 & +0 \end{array}$$

The quotient is $x^3 - 2x^2 + 4x - 8$.

D

37. Is $x - 5$ a factor of $x^5 - 19x^3 + 20x^2 - 19x + 5$?

$$\begin{array}{r|rrrrrr} 5 & 1 & -0 & -19 & +20 & -19 & +5 \\ & & +5 & +25 & +30 & +250 & +1150 \\ \hline & 1 & +5 & +6 & +50 & +231 & +1155 \end{array}$$
Divide. Use synthetic division. Insert zero for the missing x^2 term. The remainder is 1155.

Since there is a remainder, $x - 5$ is not a factor of the given polynomial.

E Maintain Your Skills

Factor completly:

41. $20x^4 - 1620$
   $= 20(x^4 - 81)$
   $= 20(x^2 + 9)(x^2 - 9)$
   $= 20(x^2 + 9)(x + 3)(x - 3)$

   There is a common factor, 20
   Difference of squares.
   Again difference of squares.

45. $5x^5 + 135x^2$
   $= 5x^2(x^3 + 27)$
   $= 5x^2(x + 3)(x^2 - 3x + 9)$

   There is a common factor $5x^2$.
   Sum of cubes.

Reduce:

49. $\dfrac{k^3 + 8}{k^3 - 4k} = \dfrac{(k + 2)(k^2 - 2k + 4)}{k(k^2 - 4)}$ Factor.

   $= \dfrac{(k + 2)(k^2 - 2k + 4)}{k(k + 2)(k - 2)}$ Divide out the common factor, $(k + 2)$.

   $= \dfrac{k^2 - 2k + 4}{k(k - 2)}$

CHAPTER 4

ROOTS AND RADICALS

EXERCISES 4.1   RATIONAL EXPONENTS

A

Simplify, using only positive exponents.

1. $144^{1/2} = 12$ The exponent, 1/2, denotes the positive square root.

5. $5^{3/5} \cdot 5^{1/5} = 5^{4/5}$ First law of exponents. When multiplying powers with the same base, add the exponents.

9. $\dfrac{x^{5/9}}{x^{2/9}} = x^{5/9 - 2/9} = x^{3/9}$ Second law of exponents. When dividing powers with the same base, subtract the exponents.

         $= x^{1/3}$ Reduce the fractional exponent.

13. $(10^{1/2})^3 = 10^{3/2}$ Third law of exponents. When raising a power to another power, multiply the exponents.

B

Simplify, using only positive exponents:

17. $3x^{1/2} \cdot 5x^{1/4} = (3 \cdot 5)(x^{1/2} \cdot x^{1/4})$ Commutative and associative laws of multiplication.

         $= 15x^{1/2 + 1/4}$ First law of exponents.

         $= 15x^{3/4}$

21. $(4^{1/3} x^{3/7})(4^{2/3} x^{3/7}) = (4^{1/3} \cdot 4^{2/3})(x^{3/7} \cdot x^{3/7})$ Commutative and associative laws of multiplication.

         $= 4^{3/3} \cdot x^{6/7}$ First law of exponents.

         $= 4x^{6/7}$ Simplify.

25. $\dfrac{-72d^{1/3}}{9d^{-1/6}} = -8d^{1/3-(-1/6)}$ Divide the coefficients.

Second law of exponents. When dividing powers with the same base, subtract the exponents.

$\qquad = -8d^{1/2}$

29. $(27x^{3/4})^{2/3} = 27^{2/3} \cdot x^{(3/4)(2/3)}$ Fourth law of exponents.

$\qquad = 9x^{1/2}$ $27^{2/3} = (27^{1/3})^2 = (3)^2$

33. $(-4c^{1/4}d^{1/2})^3 = (-4)^3 \cdot (c)^{3/4} \cdot (d)^{3/2}$ Fourth law of exponents.

$\qquad = -64c^{3/4}d^{3/2}$

37. $\left[\dfrac{x^{5/2}}{y^{2/3}}\right]^{4/5} = \dfrac{(x^{5/2})^{4/5}}{(y^{2/3})^{4/5}}$ Fifth law of exponents. When raising a quotient to a power, raise both numerator and denominator to the powers.

$\qquad = \dfrac{x^2}{y^{8/15}}$ Third law of exponents.

C

Simplify, using only positive exponents:

41. $(x^{1/2}y^2)(x^{2/3}y^{3/5}) = (x^{1/2} \cdot x^{2/3})(y^2 \cdot y^{3/5})$ Commutative and associative laws of multiplication.

$\qquad = x^{7/6}y^{13/5}$

45. $(3a^{3/5}b^{7/8})^2 = 3^2 \cdot a^{6/5} \cdot b^{14/8}$ Fourth law of exponents.

$\qquad = 9a^{6/5}b^{7/4}$

49. $(x^{-1/2}y^{2/3})^{-2/3} = (x^{-1/2})^{-2/3}(y^{2/3})^{-2/3}$ Fourth law of exponents.

$\qquad = x^{1/3}y^{-4/9}$ Third law of exponents.

$\qquad = \dfrac{x^{1/3}}{y^{4/9}}$ $y^{-4/9} = \dfrac{1}{y^{4/9}}$

53. $\left[\dfrac{x^{-5/6}}{z^{-2/3}}\right]^{-1/2} = \dfrac{(x^{-5/6})^{-1/2}}{(z^{-2/3})^{-1/2}}$ Fifth law of exponents.

$\phantom{\left[\dfrac{x^{-5/6}}{z^{-2/3}}\right]^{-1/2}} = \dfrac{x^{5/12}}{z^{1/3}}$

57. $(a^{2/3} + 3b^{1/3})(a^{1/3} - 2b^{2/3})$

$= (a^{2/3})(a^{1/3}) + (a^{2/3})(-2b^{2/3}) + (3b^{1/3})(a^{1/3}) + (3b^{1/3})(-2b^{2/3})$
 F + O + I + L

$= a^{3/3} - 2a^{2/3}b^{2/3} + 3a^{1/3}b^{1/3} - 6b^{3/3}$

$= a - 2a^{2/3}b^{2/3} + 3a^{1/3}b^{1/3} - 6b$

D

61. Two lines of a computer program read

LINE NUMBER	COMMAND
310	y = 16
320	x = y ** 1.5

What is the numerical value of x?

(The symbol "**" is used to tell the computer that the next number is an exponent.)

x = y ** 1.5	Computer statement 320.
x = 16** 1.5	Substitution, y = 16, from computer statement 310.
So x = $16^{1.5}$ = $16^{3/2}$	Algebra statement.
x = $(16^{1/2})^3$	Third law of exponents. Find the root first, then raise to the power, 3.
x = $(4)^3$	Since $16^{1/2} = 4$.
x = 64	

105

65. Two lines of a computer program read

    LINE NUMBER   COMMAND
    530   w = 16
    540   y = (w + 84) ** 1.5

What is the numerical value of y?
(Remember that the number following "**" is an exponent.)

$y = (w + 84)**1.5$	Computer statement 540.
$y = (16 + 84)**1.5$	Substitution, w = 16, from computer statement 530.
So $y = 100^{1.5} = 100^{3/2}$	Algebra statement.
$y = (100^{1/2})^3$	Third law of exponents.
$y = 10^3$	Since $100^{1/2} = 10$.
$y = 1000$	

E Maintain Your Skills

Calculate. Write answer in scientific notation:

69. $(8.46 \times 10^2)(5 \times 10^{-3})(0.2 \times 10^4)$

$= (8.46)(5)(0.2) \times (10^2)(10^{-3})(10^4)$

$= 8.46 \times 10^3$

Factor completely:

73. $84x^2 + 99x - 84$

$= 3(28x^2 + 33 - 28)$	There is a common factor, 3.
$= 3(4x + 7)(7x - 4)$	Use trial and error method.

EXERCISES 4.2 RADICALS

Recall that, unless the directions state otherwise, all variables represent only positive values.

A

Write this expression in radical form and simplify:

1. $32^{1/5} = \sqrt[5]{32}$ The denominator, 5, in the exponent indicates the fifth root.
      $= 2$ Since $2^5 = 32$.

Write this expression in radical form. Assume variables represent non-negative real numbers:

5. $(47a)^{1/2} = \sqrt{47a}$

The denominator, 2, in the exponent indicates the square root.

Write this expression with a rational exponent. Assume variables represent non-negative real numbers:

9. $\sqrt{xy} = (xy)^{1/2}$

Square root is indicated by the exponent $\frac{1}{2}$.

If the variable represents a real number, simplify. Use absolute signs where necessary.

13. $\sqrt{w^2} = |w|$

The absolute value is necessary, since w may be negative and the radical indicates only the positive square root.

B

Write this expression in radical form:

17. $6x^{1/2} = 6\sqrt{x}$

The exponent applies only to the variable x.

Write using ratitonal exponents. Assume variables represent non-negative real numbers:

21. $\sqrt{5x} = (5xy)^{1/2}$

Square root is indicated by the exponent $\frac{1}{2}$.

Find the numerical value of each expression. Use a calculator with the radical key, $\boxed{\sqrt{}}$, or a square root table (see Appendix III).

25. $\sqrt{361}$

	ENTER		DISPLAY
Using a calculator:	$\boxed{361}$ $\boxed{\sqrt{}}$		19

So $\sqrt{361} = 19$

29. $-\sqrt[4]{81}$

	ENTER		DISPLAY
Using a calculator:	$\boxed{81}$ $\boxed{\sqrt{}}$ $\boxed{\sqrt{}}$ $\boxed{+/-}$		-3

So $-\sqrt[4]{81} = -3$

33. $\sqrt[15]{1}$

	ENTER		DISPLAY
Using a calculator:	$\boxed{1}$ $\boxed{y^x}$ $\boxed{15}$ $\boxed{1/x}$ $\boxed{=}$		1

So $\sqrt[15]{1} = 1$

If the variable represents a real number, simplify. Use absolute value signs where necessary:

37. $\sqrt{4t^2} = 2|t|$ The square root of 4 is 2, and the positive square root of x^2 is $|x|$. The radical indicates the positive square root, but since x may be negative, the absolute value signs are necessary.

C

Simplify. Assume variables represent non-negative real numbers:

41. $\sqrt{121x^4} = 11x^2$ $(11x^2)^2 = 121x^4$

Write in radical form. Assume variables represent non-negative real numbers:

45. $(w^2 + 3)^{1/3} = \sqrt[3]{w^2 + 3}$ The exponent, $\frac{1}{3}$, indicates the cube root of the quantity $(w^2 + 3)$.

49. $(p - q)^{-1/2} = \dfrac{1}{(p - q)^{1/2}}$ The negative exponent indicates the reciprocal of the expression $(p - q)^{1/2}$.

$\phantom{(p-q)^{-1/2}} = \dfrac{1}{\sqrt{p - q}}$ The denominator, 2, in the exponent indicates the square root.

Write using rational exponents. Assume variables represent non-negative real numbers:

53. $\sqrt{5x^3y^5} = (5xy^5)^{1/2}$ Square root is indicated by the exponent $\frac{1}{2}$.

$\phantom{\sqrt{5x^3y^5}} = 5^{1/2}x^{3/2}y^{5/2}$ Fourth law of exponents.

If the variable represents a real number, simplify. Use absolute value signs where necessary:

57. $\sqrt{(9 - x)^2} = |9 - x|$ If $x > 9$, then the quantity $9 - x$ represents a negative value. But the radical indicates the positive square root, so the absolute value signs are necessary.

D

61. Find the length of one side of a square whose area is $196m^2$.

$s = A^{1/2}$ Formula.
$s = 196^{1/2}$ Substitute $A = 196$.
$s = \sqrt{196}$
$s = 14$ Since $14^2 = 196$.

So each side of the square is 14 m.

65. A circle has an area of 2025π square inches. Find the radius. (Hint: Area of a circle = πr^2.)

$$\begin{aligned} A &= \pi r^2 \\ 2025\pi &= \pi r^2 \\ 2025 &= r^2 \\ \sqrt{2025} &= r \\ 45 &= r \end{aligned}$$

Formula.
Substitute A = 2025π.
Divide both sides by π.
Find the square root of both sides.

The radius of the circle is 45 in.

E   Maintain Your Skills

69. Reduce: $\dfrac{98a^3b^3c^4}{40ab^5c^2} = \dfrac{\cancel{2}\cdot 7\cdot 7\cdot a^2\cdot \cancel{a}\cdot b^3\cdot c^2\cdot \cancel{c^2}}{\cancel{2}\cdot 2\cdot 2\cdot 5\cdot \cancel{a}\cdot b^3\cdot b^2\cdot \cancel{c^2}}$ Factor, and divide out the common factors.

$$= \dfrac{49a^2c^2}{20b^2}$$

73. $\dfrac{x^3 - 8x^2 - 33x}{x^3 - 5x^2} \cdot \dfrac{x^2 - 25}{x^3 - 6x^2 - 55x}$

$= \dfrac{\cancel{x}(\cancel{x-11})(x+3)}{x^2\cancel{(x-5)}} \cdot \dfrac{\cancel{(x-5)}(\cancel{x+5})}{\cancel{x}(\cancel{x-11})(\cancel{x+5})}$ Factor. Divide out common factors.

$= \dfrac{x+3}{x^2}$

EXERCISES 4.3   SIMPLIFYING AND APPROXIMATING RADICALS

A

Simplify these radicals. Assume that all variables represent positive numbers:

1. $\sqrt{200} = \sqrt{2}\,\sqrt{100} = 10\sqrt{2}$ The number 200 has a perfect square factor, 100.

5. $\sqrt{27} = \sqrt{9}\,\sqrt{3} = 3\sqrt{3}$ The number 27 has a perfect square factor, 9.

9. $\sqrt{40y^2} = \sqrt{4y^2}\,\sqrt{10} = 2y\sqrt{10}$ Assume y is positive, so no restriction (absolute value) is needed.

13. $\sqrt{24st^2} = \sqrt{4t^2}\,\sqrt{6s} = 2t\sqrt{6s}$ Assume t is positive.

B

Use a calculator or table to find the approximate values of these radicals (to the nearest thousandth). Answers below were found using a calculator:

17. $\sqrt{200} \approx 14.142$

21. $\sqrt{4428} \approx 66.543$

Simplify. Assume that all variables represent positive numbers:

25. $\sqrt{150m^{12}} = \sqrt{2 \cdot 3 \cdot 5 \cdot 5 \cdot m^2 \cdot m^2 \cdot m^2 \cdot m^2 \cdot m^2 \cdot m^2}$
$= 5m^6\sqrt{6}$

29. $\sqrt{80y^7} = \sqrt{2 \cdot 2 \cdot 2 \cdot 2 \cdot 5 \cdot y^2 \cdot y^2 \cdot y^2 \cdot y}$
$= 4y^3\sqrt{5y}$

Simplify. The variables represent real numbers. Use absolute value where necessary:

33. $\sqrt{28a^6} = \sqrt{2 \cdot 2 \cdot 7 \cdot a^2 \cdot a^2 \cdot a^2}$
$= 2|a^3|\sqrt{7}$ 　　　　Use absolute value since $a^3 < 0$ when $a < 0$.

C

Simplify. Assume all variables are positive numbers:

37. $\sqrt{175a^2b^3c^4} = \sqrt{25a^2b^2c^4}\sqrt{7b} = 5abc^2\sqrt{7b}$

41. $\sqrt[3]{8x^4y^6} = \sqrt[3]{2 \cdot 2 \cdot 2 \cdot x^3 \cdot x \cdot y^3 \cdot y^3}$
$= 2xy^2\sqrt[3]{x}$

45. $\sqrt{147\ell^{10}m^{11}} = \sqrt{49\ell^{10}m^{10}}\sqrt{3m}$
$= 7\ell^5m^5\sqrt{3m}$

Simplify. The variables represent real numbers. Use absolute value where necessary:

49. $\sqrt{48c^8d^{10}} = \sqrt{16c^8d^{10}}\sqrt{3}$
    $= 4c^4|d^5|\sqrt{3}$

    c^4 is always positive even if $c < 0$, so no absolute value bars are necessary. However, when $d < 0$, d^5 is negative, so absolute value bars are necessary.

53. $\sqrt[4]{16s^4t^8} = \sqrt[4]{2^4 \cdot s^4 \cdot t^4 \cdot t^4}$
    $= 2|s|t^2$

    Use absolute value for s since s may be a negative number. No absolute value is needed for t^2 since even if $t < 0$ (negative), $t^2 > 0$.

Reduce the index of the following radicals if the variables represent only positive numbers:

57. $\sqrt[4]{25m^2n^2} = 25^{1/4}m^{2/4}n^{2/4}$ — Change the radical form to exponential form.

    $(5^2)^{1/4}m^{1/2}n^{1/2}$ — Rewrite 25 as 5^2.

    $= 5^{2/4}m^{2/4}n^{2/4}$ — Third law of exponents.

    $= 5^{1/2}m^{1/2}b^{1/2}$ — Reduce the exponents.

    $= \sqrt{5mn}$ — Change back to radical form.

61. $\sqrt[8]{10^2w^2} = 10^{2/8}w^{2/8}$ — Change to exponential form.

    $= 10^{1/4}w^{1/4}$ — Reduce the exponents.

    $= \sqrt[4]{10w}$ — Change to radical form.

D

65. The Nice and Gooey Frosting Company advertises that one can of frosting will cover 500 square centimeters. Whatt is the length of a side of the largest square cake that can be frosted if only the top is frosted? Give answer:
a. in simplest radical form
b. to the nearest tenth of a centimeter.

Formula:
$s = \sqrt{A}$

Substitute:
$s = \sqrt{500}$ $\qquad\qquad\qquad\qquad$ A = 500

Solve:
$s = \sqrt{100}\ \sqrt{5}$
$s = 10\sqrt{5}$ $\qquad\qquad\qquad\qquad$ Simplest radical form.
$s = 22.4$ $\qquad\qquad\qquad\qquad\ \ $ Round to the nearest tenth.

Answer:
The length of cake is

a. $10\sqrt{5}$ cm in simplest radical form
b. 22.4 cm to the nearest tenth .

69. If the surface of a sphere is given, the radius of the sphere can be approximated by the formula $r = \frac{1}{2}\sqrt{0.318A}$. Find the radius of a sphere with surface area 154 square inches to the nearest tenth of an inch.

$r = \frac{1}{2}\sqrt{0.318A}$ $\qquad\qquad$ Formula.

$r = \frac{1}{2}\sqrt{(0.318)(154)}$ $\qquad$ Substitute. A = 154.

$= \frac{1}{2}\sqrt{48.972}$ $\qquad\qquad\ \ $ Multiply.

$\approx \frac{1}{2}(6.998)$ $\qquad\qquad\ \ \ $ Use a calculator to find the square root.

≈ 3.499

≈ 3.5 $\qquad\qquad\qquad\quad\ $ Round to the nearest tenth.

The radius of the sphere is approximately equal to 3.5 inches.

E Maintain Your Skills

73. Reduce: $\dfrac{8x + 4}{10x^2 + 11x + 3} = \dfrac{4(2x + 1)}{(2x + 1)(5x + 3)}$ $\qquad$ Factor, and divide out common factors.

$\qquad\qquad\qquad\qquad\qquad\qquad\quad = \dfrac{4}{5x + 3}$

Add. Reduce, if possible:

77. $\dfrac{12}{2y + 3} + 7y = \dfrac{12}{2y + 3} + \dfrac{7y(2y + 3)}{2y + 3}$ Write the second term as a fraction with denominator $2y + 3$.

$= \dfrac{12 + 7y(2y + 3)}{2y + 3}$

$= \dfrac{12 + 14y^2 + 21y}{2y + 3}$

$= \dfrac{14y^2 + 21y + 12}{2y + 3}$

EXERCISES 4.4   COMBINING RADICALS

A

Combine. Assume that all variables represent positive numbers:

1. $\sqrt{3} + \sqrt{3} = (1 + 1)\sqrt{3}$ Factor using the distributive property.

$= 2\sqrt{3}$

5. $4\sqrt{2} - 3\sqrt{2} = (4 - 3)\sqrt{2}$ Factor using the distributive property.

$= 1\sqrt{2}$

$= \sqrt{2}$ Simplify.

9. $-5\sqrt{35} + 10\sqrt{35} - 8\sqrt{35} = (-5 + 10 - 8)\sqrt{35}$

$= -3\sqrt{35}$

13. $\sqrt{18} - 4\sqrt{2} = \sqrt{9 \cdot 2} - 4\sqrt{2}$

$= 3\sqrt{2} - 4\sqrt{2}$ Simplify the first radical.

$= (3 - 4)\sqrt{2}$

$= -1\sqrt{2}$ Use the distributive property

$= -\sqrt{2}$ to factor.

B

17. $\sqrt{24} - 6\sqrt{6} + \sqrt{54} = 2\sqrt{6} - 6\sqrt{6} + 3\sqrt{6}$ Simplify each radical.

$= -\sqrt{6}$ Combine like radicals.

21. $6\sqrt{3y} + 3\sqrt{3y} = (6 + 3)\sqrt{3y}$ Factor.

$= 9\sqrt{3y}$

25. $\sqrt{25x} + \sqrt{25x} = \sqrt{25}\sqrt{x} + \sqrt{25}\sqrt{x}$ Simplify each radical.
$= 5\sqrt{x} + 5\sqrt{x}$
$= 10\sqrt{x}$ Factor and simplify.

29. $\sqrt{125x^3} - x\sqrt{45x} - \sqrt{20x^3}$
$= 5x\sqrt{5x} - 3x\sqrt{5x} - 2x\sqrt{5x}$ Simplify each radical
$= (5x - 3x - 2x)\sqrt{5x}$ Factor.
$= 0\sqrt{5x}$
$= 0$

C

33. $\sqrt{98x^3} + 7x\sqrt{18x} = 7x\sqrt{2x} + 21x\sqrt{2x}$ Simplify each radical.
$= 28x\sqrt{2x}$ Combine like radicals.

37. $5y\sqrt{63y^3} + 7y\sqrt{28y^3}$
$= 5y\sqrt{9y^2}\sqrt{7y} + 7y\sqrt{4y^2}\sqrt{7y}$ Simplify each radical.
$= 5y \cdot 3y\sqrt{7y} + 7y \cdot 2y\sqrt{7y}$
$= 15y^2\sqrt{7y} + 14y^2\sqrt{7y} = 29y^2\sqrt{7y}$ Factor and simplify.

41. $2\sqrt{40m^2n} - 3\sqrt{90m^2n}$ Simplify each radical.
$= 2\sqrt{4m}\sqrt{10n} - 3\sqrt{9m^2}\sqrt{10n}$
$= 2 \cdot 2m\sqrt{10n} - 3 \cdot 3m\sqrt{10n}$
$= 4m\sqrt{10n} - 9m\sqrt{10n} = -5m\sqrt{10n}$ Factor and simplify.

45. $13\sqrt{9x^2} - \sqrt{4x^2} - 3\sqrt{121x^2}$ Simplify each radical.
$= 13 \cdot 3x - 2x - 3 \cdot 11x = 39x - 2x - 33x = 4x$

D

49. Find the perimeter of a triangle with sides $\sqrt{80}$ in., $\sqrt{125}$ in., and $\sqrt{45}$ in.
   a. in simplest radical form
   b. to the nearest tenth of an inch.

   Formula:
   $P = a + b + c$

   Substitution:
   $P = \sqrt{80} + \sqrt{125} + \sqrt{45}$ $a = \sqrt{80}$, $b = \sqrt{125}$, $c = \sqrt{45}$.

   Solve:
   $\sqrt{16}\sqrt{5} + \sqrt{25}\sqrt{5} + \sqrt{9}\sqrt{5}$ Simplify each radical.
   $= 4\sqrt{5} + 5\sqrt{5} + 3\sqrt{5}$
   $= 12\sqrt{5}$ Factor and simplify.
   ≈ 26.8 Use a calculator.

   The perimeter of the triangle is (a) $12\sqrt{5}$ in. or (b) approximately 26.8 in.

E Maintain Your Skills

Perform the indicated operations. Reduce, if possible:

53. $2x^{-1} - 3y^{-1} = \dfrac{2}{x} - \dfrac{3}{y}$ Remember that $x^{-1} = \dfrac{1}{x}$ and $y^{-1} = \dfrac{1}{y}$.

    $= \dfrac{2y}{xy} - \dfrac{3x}{xy}$ Write each fraction with the common denominator, xy.

    $= \dfrac{2y - 3x}{xy}$

57. $\dfrac{1}{4} - \dfrac{1}{x} + \dfrac{x + 2}{x - 2}$ The LCD is $4x(x - 2)$.

    $= \dfrac{1}{4} \cdot \dfrac{x(x - 2)}{x(x - 2)} - \dfrac{1}{x} \cdot \dfrac{4(x - 2)}{4(x - 2)} + \dfrac{x + 2}{x - 2} \cdot \dfrac{4x}{4x}$

    $= \dfrac{x^2 - 2x}{4x(x - 2)} - \dfrac{4x - 8}{4x(x - 2)} + \dfrac{4x^2 + 8x}{4x(x - 2)}$

    $= \dfrac{x^2 - 2x - (4x - 8) + 4x^2 + 8x}{4x(x - 2)}$

    $= \dfrac{x^2 - 2x - 4x + 8 + 4x^2 + 8x}{4x(x - 2)}$

    $= \dfrac{5x^2 + 2x + 8}{4x(x - 2)}$

61. $\dfrac{x}{x-5} - \dfrac{3}{x+4}$

$= \dfrac{x}{x-5} \cdot \dfrac{x+4}{x+4} - \dfrac{3}{x+4} \cdot \dfrac{x-5}{x-5}$ The LCD is $(x-5)(x+4)$.

$= \dfrac{x(x+5) - 3(x-5)}{(x-5)(x+4)}$

$= \dfrac{x^2 + 4x - 3x + 15}{(x-5)(x+4)}$

$= \dfrac{x^2 + x + 15}{(x-5)(x+4)}$

EXERCISES 4.5   MULTIPLYING RADICALS

A

Multiply and simplify. All variables represent positive numbers:

1. $\sqrt{2}\,\sqrt{8} = \sqrt{16} = 4$ Multiply, then simplify.

5. $2\sqrt{6}\,\sqrt{10} = 2\sqrt{60}$ Multiply.
   $= 2\sqrt{4 \cdot 15}$ Simplify.
   $= 4\sqrt{15}$

9. $\sqrt[3]{2x}\,\sqrt[3]{5x} = \sqrt[3]{10x^2}$ Multiply. Simplest radical form, since there are no cubes in the radicand.

13. $(5\sqrt{x})(\sqrt{5x}) = 5(\sqrt{x} \cdot \sqrt{5x})$
    $= 5\sqrt{5x^2}$ Multiply.
    $= 5x\sqrt{5}$ Simplify.

B

17. $(3\sqrt{y})(2\sqrt{10}) = (3 \cdot 2)(\sqrt{y} \cdot \sqrt{10})$
    $= 6\sqrt{10y}$ Multiply.

21. $(6s\sqrt{5t})(s\sqrt{10}) = (6s \cdot s)(\sqrt{5t} \cdot \sqrt{10})$
    $= 6s^2\sqrt{50t}$ Multiply.
    $= 6s^2\sqrt{25}\,\sqrt{2t}$ Simplify.
    $= 6s^2 \cdot 5\sqrt{2t} = 30s^2\sqrt{2t}$

25. $\left[5\sqrt{10t^2}\right]\left[\sqrt{6t^3}\right] = 5\sqrt{60t^5}$ Multiply.
$\phantom{25. \left[5\sqrt{10t^2}\right]\left[\sqrt{6t^3}\right]} = 5\sqrt{4t^4}\ \sqrt{15t}$ Simplify.
$\phantom{25. \left[5\sqrt{10t^2}\right]\left[\sqrt{6t^3}\right]} = 5 \cdot 2t^2\sqrt{15t} = 10t^2\sqrt{15t}$

29. $\sqrt{3}(\sqrt{27} - \sqrt{12}) = \sqrt{81} - \sqrt{36}$ Multiply and simplify.
$\phantom{29. \sqrt{3}(\sqrt{27} - \sqrt{12})} = 9 - 6$
$\phantom{29. \sqrt{3}(\sqrt{27} - \sqrt{12})} = 3$

C

33. $(\sqrt{10} + \sqrt{2})(\sqrt{10} + \sqrt{2}) = \sqrt{100} + 2\sqrt{20} + \sqrt{4}$
$\phantom{33. (\sqrt{10} + \sqrt{2})(\sqrt{10} + \sqrt{2})} = 10 + 4\sqrt{5} + 2$
$\phantom{33. (\sqrt{10} + \sqrt{2})(\sqrt{10} + \sqrt{2})} = 12 + 4\sqrt{5}$

37. $(2\sqrt{x} + 1)(\sqrt{x} - 4) = 2\sqrt{x^2} - 7\sqrt{x} - 4$
$\phantom{37. (2\sqrt{x} + 1)(\sqrt{x} - 4)} = 2x - 7\sqrt{x} - 4$

41. $(2\sqrt{b} - \sqrt{c})(2\sqrt{b} + \sqrt{c})$ The factors are conjugate pairs.
$ = 4\sqrt{b^2} - \sqrt{c^2}$
$ = 4b - c$

45. $2\sqrt{3} \cdot 5\sqrt[4]{2} = 2 \cdot 3^{1/2} \cdot 5 \cdot 2^{1/4}$ Change to exponential form.
$\phantom{45. 2\sqrt{3} \cdot 5\sqrt[4]{2}} = 2 \cdot 5 \cdot 3^{2/4} \cdot 2^{1/4}$ Express exponents with a common denominator.
$\phantom{45. 2\sqrt{3} \cdot 5\sqrt[4]{2}} = 10\sqrt[4]{3^2 \cdot 2^1}$ Change back to radical form.
$\phantom{45. 2\sqrt{3} \cdot 5\sqrt[4]{2}} = 10\sqrt[4]{18}$ Multiply.

49. $\sqrt{2x} \cdot \sqrt[3]{3x} \cdot \sqrt[4]{4x}$ Change to exponential form.
$ = (2x)^{1/2} \cdot (3x)^{1/3} \cdot (4x)^{1/4}$
$ = (2x)^{6/12} \cdot (3x)^{4/12} \cdot (4x)^{3/12}$ Express each exponent with a common denominator.
$ = \sqrt[12]{(2x)^6} \cdot \sqrt[12]{(3x)^4} \cdot \sqrt[12]{(4x)^3}$ Change back to radical form.
$ = \sqrt[12]{64x^6} \cdot \sqrt[12]{81x^4} \cdot \sqrt[12]{64x^3}$ Simplify each radicand.
$ = \sqrt[12]{2^{12} 3^4 x^{13}} = \sqrt[12]{2^{12} x^{12}}\ \sqrt[12]{3^4 \cdot x}$ Multiply and factor.
$ = 2x\sqrt[12]{81x}$ Simplify.

D

53. Find the area of a rectangle with length $\sqrt{54}$ cm and width $\sqrt{18}$ cm. Give the answer
   a. in simplified radical form.
   b. to the nearest tenth of a square centimeter.

$A = \ell w$	Formula.
$A = \sqrt{54} \cdot \sqrt{18}$	Substitute. $\ell = \sqrt{54}$ and
$ = \sqrt{972}$	$w = \sqrt{18}$.
$ = \sqrt{2 \cdot 2 \cdot 3 \cdot 3 \cdot 3 \cdot 3 \cdot 3}$	Simplify.
$ = 2 \cdot 3 \cdot 3\sqrt{3}$	
$ = 18\sqrt{3}$	Simplified radical form.
$ = 31.2$	Rounded to nearestt tenth.

   The area is
   a. $18\sqrt{3}$ cm² in simplified radical form.
   b. 31.2 cm² to the nearest tenth.

57. Find the area of a trapezoid with bases of $\sqrt{45}$ dm and $\sqrt{80}$ dm and height $\sqrt{15}$ dm. Given the answer
   a. in simplified radical form.
   b. to the nearest tenth of a square decimeter.

   The formula for the area of a trapezoid is $A = \frac{1}{2}h(b_1 + b_2)$.

$A = \frac{1}{2}h(b_1 + b_2)$	Formula.
$A = \frac{1}{2}(\sqrt{15})(\sqrt{45} + \sqrt{80})$	Substitute. $h = \sqrt{15}$,
	$b_1 = \sqrt{45}$, and $b_2 = \sqrt{80}$.
$ = \frac{1}{2}(\sqrt{675} + \sqrt{1200})$	Multiply.
$ = \frac{1}{2}(\sqrt{225 \cdot 3} + \sqrt{400 \cdot 3})$	
$ = \frac{1}{2}(15\sqrt{3} + 20\sqrt{3})$	
$ = \frac{1}{2}(35\sqrt{3})$	
$ = \frac{35\sqrt{3}}{2}$	Simpliest radical form.
$ = 30.3$	Use a calculator, and round to the nearest tenth.

   The area of the trapezoid is
   a. $\frac{35\sqrt{3}}{2}$ dm² in simplified radical form.
   b. 30.3 dm² to the nearest tenth.

E Maintain Your Skills

Subtract. Write result with positive exponents and reduce, if possible:

61. $\dfrac{2}{a^2 + 10a - 24} - \dfrac{1}{a^2 + 17a + 60}$

$= \dfrac{2}{(a + 12)(a - 2)} - \dfrac{1}{(a + 5)(a + 12)}$

$= \dfrac{2(a + 5)}{(a + 12)(a - 2)(a + 5)} - \dfrac{1(a - 2)}{(a + 5)(a + 12)(a - 2)}$

The LCD is $(a + 12)(a - 2)(a + 5)$.

$= \dfrac{2a + 10 - (a - 2)}{(a - 2)(a + 5)(a + 12)}$

$= \dfrac{2a + 10 - a + 2}{(a - 2)(a + 5)(a + 12)}$

$= \dfrac{\cancel{a + 12}}{(a - 2)(a + 5)\cancel{(a + 12)}} = \dfrac{1}{(a - 2)(a + 5)}$

65. $\dfrac{x^{-1} + 2x}{3x - x^{-1}} = \dfrac{\frac{1}{x} + 2x}{3x - \frac{1}{x}}$ Recall that $x^{-1} = \dfrac{1}{x}$.

$= \dfrac{\frac{1}{x} + 2x}{3x - \frac{1}{x}} \cdot \dfrac{x}{x}$ Multiply both numerator and denominator by the LCD, x.

$= \dfrac{1 + 2x^2}{3x^2 - 1}$

or $\dfrac{2x^2 + 1}{3x^2 - 1}$

EXERCISES 4.6 DIVIDING RADICALS

A

Rationalize the denominator:

1. $\dfrac{2}{\sqrt{5}} = \dfrac{2}{\sqrt{5}} \cdot \dfrac{\sqrt{5}}{\sqrt{5}} = \dfrac{2\sqrt{5}}{5}$ Multiply both numerator and denominator by $\sqrt{5}$.

5. $\sqrt{\dfrac{1}{8}} = \dfrac{\sqrt{1}}{\sqrt{8}} \cdot \dfrac{\sqrt{2}}{\sqrt{2}} = \dfrac{\sqrt{2}}{4}$ Multiply numerator and denominator by $\sqrt{2}$, since $\sqrt{16} = 4$.

9. $\dfrac{-3}{\sqrt{6}} = \dfrac{-3}{\sqrt{6}} \cdot \dfrac{\sqrt{6}}{\sqrt{6}} = \dfrac{-3\sqrt{6}}{6} = -\dfrac{\sqrt{6}}{2}$ Multiply numerator and denominator by $\sqrt{6}$. Reduce.

13. $\sqrt{\dfrac{3}{8}} = \dfrac{\sqrt{3}}{\sqrt{8}} \cdot \dfrac{\sqrt{2}}{\sqrt{2}} = \dfrac{\sqrt{6}}{4}$ Multiply numerator and denominator by $\sqrt{2}$.

B

17. $\sqrt{\dfrac{a}{2}} = \dfrac{\sqrt{a}}{\sqrt{2}} \cdot \dfrac{\sqrt{2}}{\sqrt{2}} = \dfrac{\sqrt{2a}}{2}$ Multiply numerator and denominator by $\sqrt{2}$.

21. $\dfrac{2\sqrt{5}}{5\sqrt{2}} = \dfrac{2\sqrt{5}}{5\sqrt{2}} \cdot \dfrac{\sqrt{2}}{\sqrt{2}} = \dfrac{2\sqrt{10}}{5 \cdot 2}$ Multiply numerator and denominator by $\sqrt{2}$.

$\qquad\qquad\qquad = \dfrac{2\sqrt{10}}{10} = \dfrac{\sqrt{10}}{5}$ Reduce.

25. $\dfrac{-2\sqrt{8}}{\sqrt{6}} = \dfrac{-2\sqrt{8}}{\sqrt{6}} \cdot \dfrac{\sqrt{6}}{\sqrt{6}}$ Multiply numerator and denominator by $\sqrt{6}$.

$= \dfrac{-2\sqrt{48}}{6}$

$= \dfrac{-2 \cdot 4\sqrt{3}}{6}$ Simplify the radical.

$= \dfrac{-8\sqrt{3}}{6} = -\dfrac{4\sqrt{3}}{3}$ Reduce.

or

$\dfrac{-2\sqrt{8}}{\sqrt{6}} = \dfrac{-4\sqrt{2}}{\sqrt{6}} \cdot \dfrac{\sqrt{6}}{\sqrt{6}}$ Simplify the radical in the numerator, ($\sqrt{8} = 2\sqrt{2}$) and then multiply numerator and denominator by $\sqrt{6}$.

$= \dfrac{-4\sqrt{12}}{6}$

$= \dfrac{-8\sqrt{3}}{6} = -\dfrac{4\sqrt{3}}{3}$

29. $\sqrt[3]{\dfrac{3}{4}} = \dfrac{\sqrt[3]{3}}{\sqrt[3]{4}} \cdot \dfrac{\sqrt[3]{2}}{\sqrt[3]{2}}$ Multiply by the smallest radical factor that will result in a perfect cube, $\sqrt[3]{2}$.

$= \dfrac{\sqrt[3]{6}}{\sqrt[3]{8}} = \dfrac{\sqrt[3]{6}}{2}$

33. $\dfrac{2}{\sqrt[3]{25}} = \dfrac{2}{\sqrt[3]{25}} \cdot \dfrac{\sqrt[3]{5}}{\sqrt[3]{5}}$

$= \dfrac{2\sqrt[3]{5}}{\sqrt[3]{125}} = \dfrac{2\sqrt[3]{5}}{5}$

C

Rationalize the numerator:

37. $\dfrac{\sqrt{ab}}{\sqrt{3}} = \dfrac{\sqrt{ab}}{\sqrt{3}} \cdot \dfrac{\sqrt{ab}}{\sqrt{ab}}$

$= \dfrac{ab}{\sqrt{3ab}}$

Rationalize the denominator:

41. $\dfrac{\sqrt{5} - \sqrt{2}}{\sqrt{5} + \sqrt{2}}$ 　　　　　Multiply by the conjugate of the denominator to eliminate the radical.

$= \dfrac{\sqrt{5} - \sqrt{2}}{\sqrt{5} + \sqrt{2}} \cdot \dfrac{\sqrt{5} - \sqrt{2}}{\sqrt{5} - \sqrt{2}}$

$= \dfrac{5 - 2\sqrt{10} + 2}{5 - \sqrt{10} + \sqrt{10} - 2}$

$= \dfrac{7 - 2\sqrt{10}}{3}$ 　　　　　Simplify.

45. $\dfrac{2\sqrt{x}}{\sqrt{x} - \sqrt{y}}$

$= \dfrac{2\sqrt{x}}{\sqrt{x} - \sqrt{y}} \cdot \dfrac{\sqrt{x} + \sqrt{y}}{\sqrt{x} + \sqrt{y}}$ 　　　　　Multiply by the conjugate of the denominator.

$= \dfrac{2x + 2\sqrt{xy}}{x + \sqrt{xy} - \sqrt{xy} - y}$

$= \dfrac{2x + 2\sqrt{xy}}{x - y}$

49. $\dfrac{2\sqrt{5} - 3}{3\sqrt{5} + 3} = \dfrac{(2\sqrt{5} - 3)}{(3\sqrt{5} + 3)} \cdot \dfrac{(3\sqrt{5} - 3)}{(3\sqrt{5} - 3)}$ 　　　　　Multiply by the conjugate.

$= \dfrac{6\sqrt{25} - 6\sqrt{5} - 9\sqrt{5} + 9}{9\sqrt{25} - 9}$

$= \dfrac{(30 + 9) + (-6\sqrt{5} - 9\sqrt{5})}{45 - 9}$ 　　　　　Simplify and combine like terms.

$= \dfrac{39 - 15\sqrt{5}}{36}$

$= \dfrac{\cancel{3}(13 - 5\sqrt{5})}{\cancel{3}(12)}$ 　　　　　Divide out the common factor, 3.

$= \dfrac{13 - 5\sqrt{5}}{12}$

53. $\dfrac{3\sqrt{x} - \sqrt{y}}{5\sqrt{x} + 2\sqrt{y}}$

$\qquad = \dfrac{3\sqrt{x} - \sqrt{y}}{5\sqrt{x} + 2\sqrt{y}} \cdot \dfrac{5\sqrt{x} - 2\sqrt{y}}{5\sqrt{x} - 2\sqrt{y}}$ Multiply by the conjugate.

$\qquad = \dfrac{15x - 11\sqrt{xy} + 2y}{25x - 4y}$ Simplify.

Rationalize the numerator:

57. $\dfrac{2 + \sqrt{3}}{1 + \sqrt{3}} = \dfrac{2 + \sqrt{3}}{1 + \sqrt{3}} \cdot \dfrac{2 - \sqrt{3}}{2 - \sqrt{3}}$ Multiply by the conjugate of the numerator.

$\qquad = \dfrac{4 - 2\sqrt{3} + 2\sqrt{3} - 3}{2 - \sqrt{3} + 2\sqrt{3} - 3}$

$\qquad = \dfrac{1}{-1 + \sqrt{3}}$ Simplify.

D

61. A pendulum is 4 inches $\left(\dfrac{1}{3} \text{ foot}\right)$ long. Find the time it takes to complete one cycle. Give the answer
 a. in simplified radical form.
 b. to the nearest tenth of a second.

$\qquad\qquad\qquad\qquad\qquad$ Use the formula from the application problem given in the text.

$T = 2\pi \dfrac{\sqrt{L}}{\sqrt{32}}$ Formula. L represents the length of the pendulum in feet, and T represents the time in seconds.

$T = 2\pi \dfrac{\sqrt{\tfrac{1}{3}}}{\sqrt{32}}$ Substitute. $L = \dfrac{1}{3}$.

$T = 2\pi \dfrac{\sqrt{\tfrac{1}{3}}}{\sqrt{32}} \cdot \dfrac{\sqrt{2}}{\sqrt{2}}$ Rationalize the denominator.

$\quad = 2\pi \dfrac{\sqrt{\tfrac{2}{3}}}{8}$

$= 2\pi \dfrac{\sqrt{\dfrac{2}{3} \cdot \dfrac{3}{3}}}{8}$ Simplify.

$= 2\pi \cdot \dfrac{\dfrac{\sqrt{6}}{3}}{8}$ $\dfrac{\sqrt{6}}{3} \div 3 = \dfrac{\sqrt{6}}{3} \cdot \dfrac{1}{8}$

$= 2\pi \cdot \dfrac{\sqrt{6}}{24}$

$= \dfrac{\pi\sqrt{6}}{12}$ Simplest radical form.

≈ 0.6 Use a calculator and round to the nearest tenth.

The time it takes the pendulum to complete one cycle is

a. $\dfrac{\pi\sqrt{6}}{12}$ sec in simplified radical form.

b. 0.6 sec to the nearest tenth.

E Maintain Your Skills

Subtract. Write results with positive exponents and reduce, if possible:

65. $\dfrac{6b}{1 - b^2} - \dfrac{3}{b - 1}$

$= \dfrac{6b}{1 - b^2} - \dfrac{3}{b - 1} \cdot \dfrac{-1}{-1}$ Multiply by $\dfrac{-1}{-1}$.

$= \dfrac{6b}{1 - b^2} - \dfrac{-3}{1 - b}$ Now, the LCD is $1 - b^2$, or $(1 - b)(1 + b)$.

$= \dfrac{6b}{(1 - b)(1 + b)} - \dfrac{-3(1 + b)}{(1 - b)(1 + b)}$ Write both fractions with a common denominator.

$= \dfrac{6b}{(1 - b)(1 + b)} - \dfrac{-3 - 3b}{(1 - b)(1 + b)}$

$= \dfrac{6b - (-3 - 3b)}{(1 - b)(1 + b)}$

$= \dfrac{6b + 3 + 3b}{(1 - b)(1 + b)}$

$= \dfrac{9b + 3}{(1 - b)(1 + b)}$

Simplify:

69. $\dfrac{2 + \dfrac{3}{xy}}{\dfrac{4}{xy} - 1} = \dfrac{2 + \dfrac{3}{xy}}{\dfrac{4}{xy} - 1} \cdot \dfrac{xy}{xy}$ Multiply both numerator and denominator by the LCD, xy.

$= \dfrac{2xy + 3}{4 - xy}$

EXERCISES 4.7 COMPLEX NUMBERS

A

Write in standard form:

1. $\sqrt{-16} = \sqrt{16 \cdot -1} = \sqrt{16}\,\sqrt{-1}$
   $= 4\sqrt{-1}$ Since $\sqrt{16} = 4$
   $= 4i$ Since $\sqrt{-1} = i$.

5. $\sqrt{-18} = \sqrt{9 \cdot 2 \cdot -1}$
   $= \sqrt{9} \cdot \sqrt{-1} \cdot \sqrt{2}$
   $= 3i\sqrt{2}$

9. $\sqrt{-75} - \sqrt{-3} = \sqrt{75 \cdot -1} - \sqrt{3 \cdot -1}$
   $= 5i\sqrt{3} - i\sqrt{3}$ Since $\sqrt{75} = 5\sqrt{3}$, and $\sqrt{-1} = i$.
   $= 4i\sqrt{3}$

Simplify:

13. $i^{54} = i^{52} \cdot i^2$ Factor: 52 is the largest multiple of four that is less than 54.
    $= i^{4 \cdot 13} \cdot i^2$
    $= (i^4)^{13} \cdot i^2$
    $= 1^{13} \cdot i^2$ $i^4 = 1$.
    $= -1$ Since $i^2 = -1$.

Add or subtract:

17. $(4 + 2i) + (3 + 4i)$
    $= (4 + 3) + (2i + 4i)$ Regroup the real parts and imaginary parts of the two complex numbers.
    $= 7 + 6i$

126

21. $(4 + \sqrt{-1} + 2 + \sqrt{-9})$
    $= (4 + i) + (2 + 3i)$ Write each complex number in standard form.
    $= (4 + 2) + (i + 3i)$ Regroup.
    $= 6 + 4i$

B

Perform the indicated operations:

25. $7 - (-8 - i) = 7 + (8 + i)$ Subtract by adding the
    $= 15 + i$ opposite.

29. $(8 + 3i) - (-3 + i) = (8 + 3i) + (3 - i)$
    $= (8 + 3) + (3i - i)$
    $= 11 + 2i$

33. $3i(2 + 5i) = 6i + 15i^2$ When written in standard form, complex numbers can be multiplied in the same way as polynomials.
    $= -15 + 6i$ Since $i^2 = -1$.

37. $\dfrac{i^{16}}{i^{21}} = \dfrac{1}{i^5}$ Use laws of exponents to reduce the fraction.

    $= \dfrac{1}{i^5} \cdot \dfrac{i^3}{i^3} = \dfrac{i^3}{i^8}$ Multiply by one, $\dfrac{i^3}{i^3}$, so that the denominator will be a power of i that is a multiple of 4.

    $= \dfrac{i^3}{(i^4)^2} = \dfrac{i^3}{1^2} = i^3 = -i$ Since $i^3 = -i$.

41. $\dfrac{i^{21}}{i^{-10}} = i^{31}$ Second law of exponents.

    $= i^{28} \cdot i^3$ The largest multiple of four that is less than 31 is 28.
    $= (i^4)^7 \cdot i^3$
    $= 1^7 \cdot i^3$ $i^4 = 1$
    $= i^3$
    $= -i$ $i^3 = i^2 \cdot i = -1 \cdot i = -i$.

C

Perform the indicated operations:

45. $(7 + \sqrt{-16})(-2 - \sqrt{-25})$
    $= (7 + 4i)(-2 - 5i)$
    $= -14 - 35i - 8i - 20i^2$ In standard form, complex numbers can be multiplied like polynomials. Substitute -1 for i^2.

    $= -14 - 43i - 20(-1)$
    $= -14 - 43i + 20 = 6 - 43i$

49. $\dfrac{1}{2i} = \dfrac{1}{2i} \cdot \dfrac{i^3}{i^3} = \dfrac{i^3}{2i^4}$ Multiply by one so that the denominator will be a power of i that is a multiple of 4.

    $= \dfrac{-i}{2(1)} = -\dfrac{i}{2}$ or $-\dfrac{1}{2}i$ Since $i^3 = -i$ and $i^4 = 1$.

53. $(\sqrt{8} + i\sqrt{12})(\sqrt{2} - i\sqrt{3})$
    $= \sqrt{16} - i\sqrt{24} + i\sqrt{24} - i^2\sqrt{36}$ Multiply.
    $= 4 - 6i$ Simplify.
    $= 4 - 6(-1) = 4 + 6 = 10$ Replace i^2 with -1.

57. $\dfrac{-5 - 4i}{-2 + 3i} = \dfrac{(-5 - 4i)}{(-2 + 3i)} \cdot \dfrac{(-2 - 3i)}{(-2 - 3i)}$ Multiply numerator and denominator by the conjugate of the denominator.

    $= \dfrac{10 + 15i + 8i + 12i^2}{4 - 9i^2}$

    $= \dfrac{10 + 23i + 12(-1)}{4 - 9(-1)}$ Replace i^2 with -1, and combine like terms.

    $= \dfrac{-2 + 23i}{13}$ or $-\dfrac{2}{13} + \dfrac{23}{13}i$ Standard form.

61. $\dfrac{3}{(2 - i)^2} = \dfrac{3}{4 - 4i + i^2} = \dfrac{3}{4 - 4i - 1}$ Multiply and simplify, in the denominator.

    $= \dfrac{3}{3 - 4i} \cdot \dfrac{3 + 4i}{3 + 4i}$ Multiply by the conjugate of the denominator.

    $= \dfrac{9 + 12i}{9 - 16i^2}$

    $= \dfrac{9 + 12i}{9 + 16} = \dfrac{9 + 12i}{25}$ Fraction form.

    $= \dfrac{9}{25} + \dfrac{12}{25}i$ Standard form.

65. $\dfrac{-6i}{3+\sqrt{-3}} = \dfrac{-6i}{3+i\sqrt{3}}$ Simplify in the denominator.

$= \dfrac{-6i}{3+i\sqrt{3}} \cdot \dfrac{3-i\sqrt{3}}{3-i\sqrt{3}}$ Multiply by the conjugate.

$= \dfrac{-18i + 6\sqrt{3}i^2}{9 - 3i^2}$

$= \dfrac{-18i - 6\sqrt{3}}{9 + 3}$ $i^2 = -1$.

$= \dfrac{-6\sqrt{3} - 18i}{12}$

$= \dfrac{-\sqrt{3} - 3i}{2}$ Reduced fraction form.

$= -\dfrac{\sqrt{3}}{2} - \dfrac{3}{2}i$ Standard form.

D

A formula to find the total series of impedance is $Z_T = Z_1 + Z_2$ where Z_T represents the total impedance, $Z_1 = R_1 + jX_{C_1}$ and $Z_2 = R_2 + jX_{C_2}$. Find the total series impedance of resistance and capacitance if:

69. $R_1 = 15700\Omega$, $X_{C_1} = 18900\Omega$, $R_2 = 23800\Omega$, and $X_{C_2} = 30500\Omega$.

Formulas:
$Z_T = Z_1 + Z_2$
$Z_1 = R_1 + jX_{C_2}$
$Z_2 = R_2 + jX_{C_2}$

Substitute:
$Z_T = (15700 + j18900) + (23800 + j30500)$

Solve:
$Z_T = 39500 + j49400$

The total impedance is $39500 + j49400$ ohms.

E Maintain Your Skills

Simplify:

73. $1 + \dfrac{1}{1 + \dfrac{1}{1 + x}}$

$= 1 + \dfrac{1}{1 + \dfrac{1}{1 + x}} \cdot \dfrac{1 + x}{1 + x}$ Multiply both numerator and denominator by $1 + x$.

$= 1 + \dfrac{1}{1 + x + 1}$ Use distributive property in denominator.

$= \dfrac{1}{1} + \dfrac{1 + x}{x + 2}$ Combine like terms in denominator.

$= \dfrac{1(x + 2)}{1(x + 2)} + \dfrac{1 + x}{x + 2}$ Build so that $x + z$ is the common denominator.

$= \dfrac{x + 2 + 1 + x}{x + 2} = \dfrac{2x + 3}{x + 2}$ Simplify.

Divide:

77. $2x - 5 \overline{) 2x^3 - 9x^2 + 25}$

$\begin{array}{r} x^2 - 2x - 5 \\ 2x - 5 \overline{) 2x^3 - 9x^2 + 0x + 25} \\ \underline{2x^3 - 5x^2} \\ -4x^2 + 0x \\ \underline{-4x^2 + 10x} \\ -10x + 25 \\ \underline{-10x + 25} \\ 0 \end{array}$ Insert 0 for the missing x term.

The quotient is $x^2 - 2x - 5$.

CHAPTER 5

QUADRATIC EQUATIONS

EXERCISES 5.1 QUADRATIC EQUATIONS SOLVED BY FACTORING

A

Write each of the following quadratic equations in standard form and identify a, b, and c:

1. $4x^2 - 5x = 13x + 17$

   $4x^2 - 18x - 17 = 0$ The standard form of a quadratic equation is $ax^2 + bx + c = 0$.

   $a = 4$, $b = -18$, $c = -17$

Solve:

5. $2x(x - 5) = 0$
   $2x = 0$ or $x - 5 = 0$ Zero-product property.
   $x = 0$ or $x = 5$

   The solution set is $\{0, 5\}$.

9. $(2x + 5)(3x - 1) = 0$
   $2x + 5 = 0$ or $3x - 1 = 0$ Zero-product property.
   $2x = -5$ $3x = 1$
   $x = -\frac{5}{2}$ or $x = \frac{1}{3}$

   The solution set is $\left\{-\frac{5}{2}, \frac{1}{3}\right\}$.

13. $a^2 + 7a + 10 = 0$
    $(a + 5)(a + 2) = 0$ Factor.
    $a + 5 = 0$ or $a + 2 = 0$ Zero-product law.
    $a = -5$ $a = -2$

    The solution set is $\{-5, -2\}$.

B

17. $y^2 + 6y - 72 = 0$
    $(y + 12)(y - 6) = 0$ Factor.
    $y + 12 = 0$ or $y - 6 = 0$ Zero-product law.
    $y = -12$ $y = 6$

    The solution set is $\{-12, 6\}$.

21. $\quad 3x^2 + 16x + 5 = 0$
$\quad\quad (3x + 1)(x + 5) = 0$ Factor.
$\quad\quad 3x + 1 = 0 \text{ or } x + 5 = 0$
$\quad\quad\quad\quad x = -\dfrac{1}{3} \text{ or } \quad x = -5$

  The solution set is $\left\{-5, -\dfrac{1}{3}\right\}$.

25. $\quad 6x^2 + 21x + 15 = 0$
$\quad\quad\quad 2x^2 + 7x + 5 = 0$ Divide both sides by the common factor, 3.
$\quad\quad (2x + 5)(x + 1) = 0$ Factor the left side.
$\quad\quad 2x + 5 = 0 \text{ or } x + 1 = 0$ Zero-product property.
$\quad\quad\quad\quad 2x = -5$
$\quad\quad\quad\quad x = -\dfrac{5}{2} \text{ or } \quad x = -1$

  The solution set is $\left\{-\dfrac{5}{2}, -1\right\}$.

29. $\quad\quad\quad 4x^2 - 25 = 0$
$\quad\quad (2x + 5)(2x - 5) = 0$ Factor.
$\quad\quad 2x + 5 = 0 \text{ or } 2x - 5 = 0$
$\quad\quad\quad x = -\dfrac{5}{2} \quad\quad x = \dfrac{5}{2}$

  The solution set is $\left\{\pm\dfrac{5}{2}\right\}$.

C

33. $\quad 6x^2 - 11x - 10 = 0$
$\quad\quad (3x + 2)(2x - 5) = 0$
$\quad\quad 3x + 2 = 0 \text{ or } 2x - 5 = 0$
$\quad\quad\quad x = -\dfrac{2}{3} \quad\quad x = \dfrac{5}{2}$

  The solution set is $\left\{-\dfrac{2}{3}, \dfrac{5}{2}\right\}$.

37. $\quad\quad\quad 3z^2 - 22z = -7$
$\quad\quad 3z^2 - 22z + 7 = 0$ Write the equation in standard form.
$\quad\quad (3z - 1)(z - 7) = 0$ Factor.
$\quad\quad 3z - 1 = 0 \text{ or } z - 7 = 0$ Zero-product law.
$\quad\quad\quad z = \dfrac{1}{3} \quad\quad z = 7$

  The solution set is $\left\{\dfrac{1}{3}, 7\right\}$.

41.
$$3x^2 + 7 = 10x$$
$$3x^2 - 10x + 7 = 0$$ Write the equation in standard form.
$$(3x - 7)(x - 1) = 0$$ Factor.
$$3x - 7 = 0 \text{ or } x - 1 = 0$$ Zero-product property.
$$3x = 7$$
$$x = \frac{7}{3} \text{ or } x = 1$$

The solution set is $\left\{1, \frac{7}{3}\right\}$.

45.
$$4x(x - 3) = -9$$
$$4x^2 - 12x + 9 = 0$$ Multiply and write the equation in standard form.
$$(2x - 3)(2x - 3) = 0$$ Factor.
$$2x - 3 = 0 \text{ or } 2x - 3 = 0$$
$$x = \frac{3}{2} \qquad x = \frac{3}{2}$$ A double root.

The solution set is $\left\{\frac{3}{2}\right\}$.

D

49. If the profit for an electronics manufacturer is $P = x^2 - 135x$, how many circuit boards must be made before the break-even point is reached? (P = 0)

<u>Formula:</u>
$P = x^2 - 135x$

<u>Substitute:</u>
$0 = x^2 - 135x$ The break-even point is when the profit is zero, so substitute 0 for P.

<u>Solve:</u>
$0 = x(x - 135)$ Factor.
$x = 0$ or $x - 135 = 0$ Zero-product law.
$x = 0 \qquad x = 135$ Reject $x = 0$, since no (zero) boards are made.

The number of circuit boards that must be made to reach the break-even point is 135.

53. The profit P in dollars for Polore Products to manufacture and sell x koldies is given by $P = -x^2 + 20x$. How many koldies were sold if the business lost $300?

Formula:
$P = -x^2 + 20x$

Substitute:
$-300 = -x^2 + 20x$ $P = -300$ since there was a loss of $300.

Solve:
$x^2 - 20x - 300 = 0$ Write the equation in standard form.

$(x - 30)(x + 10) = 0$ Factor.
$x - 30 = 0$ or $x + 10 = 0$ Zero-product law.
$\quad x = 30 \quad\quad\quad x = -10$ Reject the negative quantity, since x represents koldies sold and must be positive.

So 30 koldies were sold.

E Maintain Your Skills

Simplify using only positive exponents:

57. $\left[\dfrac{p^{-3/8}}{p^{-1/2}}\right]^4 = \dfrac{(p^{-3/8})^4}{(p^{-1/2})^4}$ When raising a quotient to a power, raise both numerator and denominator to the power.

$= \dfrac{p^{-3/2}}{p^{-2}}$ When raising a power to a power, multiply the exponents.

$= \dfrac{1}{p^{3/2}} \cdot \dfrac{p^2}{1} = \dfrac{p^2}{p^{3/2}}$ Rewrite powers using positive exponents.

$= p^{2-3/2} = p^{1/2}$ When dividing powers, subtract the exponents.

61. $3x^{4/5}(2x^{1/5} - 5x^{2/5})$

$= 3x^{4/5}(2x^{1/5}) + 3x^{4/5}(-5x^{2/5})$ Distributive property.

$= 3 \cdot 2 \cdot x^{4/5} x^{1/5} + 3(-5) \cdot x^{4/5} \cdot x^{2/5}$ Regroup factors.

$= 6x^{5/5} - 15x^{6/5}$ First Law of Exponents.

$= 6x - 15x^{6/5}$

EXERCISES 5.2   QUADRATIC EQUATIONS SOLVED BY SQUARE ROOTS

A

Solve:

1. $x^2 = 1$
    $x = \pm\sqrt{1} = \pm 1$ If $x^2 = k$, then $x = \pm\sqrt{k}$.

    The solution set is $\{\pm 1\}$.

5. $a^2 = 0.16$
    $a = \pm\sqrt{0.16} = \pm 0.4$

    The solution set is $\{\pm 0.4\}$.

9. $a^2 = \frac{4}{9}$
    $a = \pm\sqrt{\frac{4}{9}} = \pm\frac{2}{3}$ If $x^2 = k$, then $x = \pm\sqrt{k}$.

    The solution set is $\left\{\pm\frac{2}{3}\right\}$.

13. $y^2 = 28$
    $y = \pm\sqrt{28} = \pm\sqrt{4}\,\sqrt{7}$
    $ = \pm 2\sqrt{7}$

    The solution set is $\{\pm 2\sqrt{7}\}$.

B

17. $(x + 1)^2 = 1$
    $x + 1 = \pm\sqrt{1} = \pm 1$ If $x^2 = k$, then $x = \pm\sqrt{k}$.
    $x + 1 = 1$ or $x + 1 = -1$ Rewrite as two equations.
    $x = 0 x = -2$

    The solution set is $\{-2, 0\}$.

21. $(c + 4)^2 = 0$
    $c + 4 = \pm\sqrt{0}$
    $c + 4 = 0$
    $c = -4$

    The solution set is $\{-4\}$.

135

25. $(x + 6)^2 = -5$

$\quad\quad x + 6 = \pm\sqrt{-5}$ $\quad\quad\quad\quad\quad\quad\quad$ $\sqrt{-5} = \sqrt{-1}\,\sqrt{5}$

$\quad\quad x + 6 = \pm i\sqrt{5}$ $\quad\quad\quad\quad\quad\quad\quad\quad\quad\quad = i\sqrt{5}$

$\quad\quad x + 6 = -i\sqrt{5}$ or $x + 6 = i\sqrt{5}$ $\quad$ Rewrite as two equations.

$\quad\quad\quad\quad x = -6 - i\sqrt{5}$ or $x = -6 + i\sqrt{5}$

The solution set is $\{-6 \pm i\sqrt{5}\}$.

29. $(x - 2)^2 = 48$

$\quad\quad x - 2 = \pm\sqrt{48}$ $\quad\quad\quad\quad\quad\quad\quad$ If $x^2 = k$, then $x = \pm\sqrt{k}$.

$\quad\quad x - 2 = \pm 4\sqrt{3}$

$\quad\quad x - 2 = -4\sqrt{3}$ or $\quad x - 2 = 4\sqrt{3}$

$\quad\quad\quad\quad x = 2 - 4\sqrt{3}$ or $\quad\quad x = 2 + 4\sqrt{3}$

The solution set is $\{2 \pm 4\sqrt{3}\}$.

C

33. $x^2 + 12x + 36 = 4$

$\quad\quad (x + 6)^2 = 4$ $\quad\quad\quad\quad\quad\quad\quad\quad\quad$ Rewrite the perfect square trinomial as the square of a binomial.

$\quad\quad\quad\quad x + 6 = \pm\sqrt{4} = \pm 2$

$\quad x + 6 = 2$ or $x + 6 = -2$

$\quad\quad\quad x = -4 \quad\quad\quad\quad x = -8$

The solution set is $\{-8, -4\}$.

37. $x^2 + 4x + 4 = -3$

$\quad\quad (x + 2)^2 = -3$

$\quad\quad\quad\quad x + 2 = \pm\sqrt{-3}$ $\quad\quad\quad\quad\quad\quad$ $\sqrt{-3} = \sqrt{-1}\,\sqrt{3}$

$\quad\quad\quad\quad x + 2 = \pm i\sqrt{3}$ $\quad\quad\quad\quad\quad\quad\quad\quad\quad = i\sqrt{3}$

$\quad\quad x + 2 = i\sqrt{3}$ or $x + 2 = -i\sqrt{3}$

$\quad\quad\quad\quad x = -2 + i\sqrt{3} \quad\quad x = -2 - i\sqrt{3}$

The solution set is $\{-2 \pm i\sqrt{3}\}$.

41. $x^2 + 32x + 256 = -64$

$\quad\quad (x + 16)^2 = -64$

$\quad\quad\quad\quad x + 16 = \pm\sqrt{-64}$ $\quad\quad\quad\quad\quad\quad$ $\sqrt{-64} = \sqrt{-1}\,\sqrt{64}$

$\quad\quad\quad\quad x + 16 = \pm 8i$ $\quad\quad\quad\quad\quad\quad\quad\quad\quad\quad = 8i$

$\quad x + 16 = 8i$ or $x + 16 = -8i$

$\quad\quad\quad x = -16 + 8i \quad\quad x = -16 - 8i$

The solution set is $\{-16 \pm 8i\}$.

45. $4x^2 - 28x + 49 = 40$
     $(2x - 7)^2 = 40$ Factor the left side, a
     $2x - 7 = \pm\sqrt{40}$ perfect square trinomial.
     $2x - 7 = \pm 2\sqrt{10}$
     $2x - 7 = -2\sqrt{10}$ or $2x - 7 = 2\sqrt{10}$
     $2x = 7 - 2\sqrt{10}$ or $x = 7 + 2\sqrt{10}$
     $x = \dfrac{7 + 2\sqrt{10}}{2}$ $x = \dfrac{7 + 2\sqrt{10}}{2}$

The solution set is $\left\{\dfrac{7 \pm 2\sqrt{10}}{2}\right\}$.

D

49. The illumination (in foot candles) of a light source varies inversely as the square of the distance from the light source. The formula is $Id^2 = K$. (I is the illuminaion, and d is the distance.) If the constant of variation is 36, what is the distance from the light source if the illumination is 2 foot candles?

<u>Formula:</u>
$Id^2 = K$

<u>Substitute:</u>
$2d^2 = 36$ K = 36 and I = 2.

<u>Solve:</u>
$d^2 = 18$ Divide each side by 2.
$d = \pm\sqrt{18} = \pm\sqrt{9}\sqrt{2}$
$d = \pm 3\sqrt{2}$ Reject $-3\sqrt{2}$, since distance is not measured with negative numbers.

The distance is $3\sqrt{2}$ feet, or approximaetly 4.24 feet.

Solve using the Pythagorean Theorem:

53. If a = 5 and c = 13, then b = ?

    $b^2 = c^2 - a^2$ Pythagorean Theorem.
    $b^2 = 13^2 - 5^2$ Let c = 13 and a = 5.
    $b^2 = 169 - 25 = 144$
    $b = \pm\sqrt{144} = \pm 12$ Disregard the negative root since a length cannot be negative.

    So the length of side b is 12.

57. Find the length of the border of a garden plot that is in the shape of a right triangle if the two shorter sides are 12 ft and 16 ft in length.

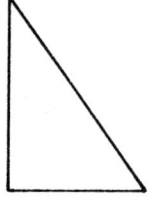

$c^2 = a^2 + b^2$ Pythagorean Theorem.
$c^2 = 12^2 + 16^2$ Let a = 12, b = 16.
$c^2 = 400$
$c = \pm\sqrt{400} = \pm 20$ Disregard the negative root since a length cannot be negative.

The length of the third side is 20 ft.

E   Maintain Your Skills

Write in radical form. Simplify, if possible:

61. $8^{-4/3} = \dfrac{1}{8^{4/3}}$ Definition of a negative exponent.

$= \dfrac{1}{\sqrt[3]{8^4}} \cdot \dfrac{\sqrt[3]{8^2}}{\sqrt[3]{8^2}}$ Rationalize the denominator.

$= \dfrac{4}{\sqrt[3]{8^6}}$

$= \dfrac{4}{8^2} = \dfrac{4}{64} = \dfrac{1}{16}$

Write in radical form:

65. $7t^{1/2} = 7(t^{1/2}) = 7(\sqrt{t}) = 7\sqrt{t}$

EXERCISES 5.3   QUADRATIC EQUATIONS SOLVED BY COMPLETING THE SQUARE

A

Solve by completing the square; check by factoring when possible:

1. $x^2 - 2x - 3 = 0$
   $x^2 - 2x = 3$ Add 3 to both sides.
   $x^2 - 2x + 1 = 3 + 1$ Add the square of half the coefficient of x to both sides: $\left[\frac{1}{2}(-2)\right]^2$.

   $(x - 1)^2 = 4$ Factor the left side, and simplify the right side.
   $x - 1 = \pm 2$ Solve.
   $x - 1 = 2$ or $x - 1 = -2$
   $x = 3$ or $x = -1$

   or
   $x^2 - 2x - 3 = 0$
   $(x - 3)(x + 1) = 0$ Check by factoring.
   $x - 3 = 0$ or $x + 1 = 0$
   $x = 3$ or $x = -1$ Checks.

   The solution set is $\{-1, 3\}$.

5. $x^2 - 10x - 24 = 0$
   $x^2 - 10x = 24$ Add 24 to both sides.
   $x^2 - 10x + 25 = 24 + 25$ Add the square of half the coefficient of x: $\left[\frac{1}{2}(-10)\right]^2$.

   $(x - 5)^2 = 49$ Factor on the left. Simplify on the right.
   $x - 5 = \pm 7$
   $x - 5 = 7$ or $x - 5 = -7$
   $x = 12$ or $x = -2$

   $x^2 - 10x - 24 = 0$
   $(x - 12)(x + 2) = 0$ Check by factoring.
   $x - 12 = 0$ or $x + 2 = 0$
   $x = 12$ or $x = -2$ Checks.

   The solution set is $\{-2, 12\}$.

9. $x^2 - 2x - 8 = 0$
$x^2 - 2x = 8$ Add 8 to both sides.
$x^2 - 2x + 1 = 8 + 1$ Add the square of half the coefficient of x to both sides: $\left[\frac{1}{2}(-2)\right]^2$.
                                  Factor on the left, and simplify on the right.
$(x - 1)^2 = 9$
$x - 1 = \pm 3$
$x - 1 = 3$ or $x - 1 = -3$
$x = 4$ or $x = -2$

$x^2 - 2x - 8 = 0$ Check by factoring.
$(x - 4)(x + 2) = 0$
$x = 4$ or $x = -2$ Checks.

The solution set is $\{-2, 4\}$.

13. $z^2 + 2z - 3 = 0$
$z^2 + 2z = 3$ Add 3 to both sides.
$z^2 + 2z + 1 = 3 + 1$ Add $\left[\frac{1}{2}(2)\right]^2$.
$(z + 1)^2 = 4$ Factor on the left.
$z + 1 = \pm 2$
$z + 1 = -2$ or $z + 1 = 2$
$z = -3$ or $z = 1$

$z^2 + 2z - 3 = 0$ Factor the original equation to check.
$(z + 3)(z - 1) = 0$
$z = -3$ or $z = 1$ Checks.

The solution set is $\{-3, 1\}$.

17. $x^2 - 7x + 12 = 0$
$x^2 - 7x = -12$ Subtract 12 from both sides.
$x^2 - 7x + \frac{49}{4} = -\frac{48}{4} + \frac{49}{4}$ Add the square of one-half the coefficient of x: $\left[\frac{1}{2}(-7)\right]^2$.
                                  Factor on the left and simplify on the right.
$\left[x - \frac{7}{2}\right]^2 = \frac{1}{4}$
$x - \frac{7}{2} = \pm \frac{1}{2}$
$x - \frac{7}{2} = -\frac{1}{2}$ or $x - \frac{7}{2} = \frac{1}{2}$
$x = \frac{6}{2}$ or $x = \frac{8}{2}$
$x = 3$ or $x = 4$

$x^2 - 7x + 12 = 0$ Check by factoring.
$(x - 3)(x - 4) = 0$
$x = 3$ or $x = 4$ Checks.

The solution set is $\{3, 4\}$.

140

B

Solve by completing the square. (Optional: check with calculator.)

21. $3x^2 + 12x + 6 = 0$
    $x^2 + 4x + 2 = 0$ Divide both sides by 3.
    $x^2 + 4x = -2$ Subtract 2 from both sides.
    $x^2 + 4x + 4 = -2 + 4$ Complete the square by adding $\left[\frac{1}{2}(4)\right]^2$ to both sides.

    $(x + 2)^2 = 2$ Factor the left side and simplify the right side.

    $x + 2 = \pm\sqrt{2}$ Solve.

    $x + 2 = \sqrt{2}$ or $x + 2 = -\sqrt{2}$
    $x = -2 + \sqrt{2}$ or $x = -2 - \sqrt{2}$

    The solution set is $\{-2 \pm \sqrt{2}\}$.

25. $2x^2 + 24x + 8 = 0$
    $x^2 + 12x + 4 = 0$ Divide both sides by 2.
    $x^2 + 12x = -4$ Subtract 4 from both sides.
    $x^2 + 12x + 36 = -4 + 36$ Add $\left[\frac{1}{2}(12)\right]^2$ to both sides.

    $(x + 6)^2 = 32$ Factor on the left, and simplify on the right.

    $x + 6 = \pm\sqrt{32}$ Solve.
    $x + 6 = \pm 4\sqrt{2}$

    $x + 6 = -4\sqrt{2}$ or $x + 6 = 4\sqrt{2}$
    $x = -6 - 4\sqrt{2}$ or $x = -6 + 4\sqrt{2}$

    The solution set is $\{-6 \pm 4\sqrt{2}\}$.

29. $x^2 - 3x + 5 = 0$
    $x^2 - 3x = -5$ Subtract 5 from both sides.
    $x^2 - 3x + \frac{9}{4} = -5 + \frac{9}{4}$ Add $\left[\frac{1}{2}(-3)\right]^2$ to both sides.

    $\left[x - \frac{3}{2}\right]^2 = -\frac{11}{4}$ Factor on the left, and simplify on the right.

    $x - \frac{3}{2} = \pm\sqrt{-\frac{11}{4}}$ Solve.

    $x - \frac{3}{2} = \pm\frac{\sqrt{11}}{2}i$

    $x - \frac{3}{2} = \frac{\sqrt{11}}{2}i$ or $x - \frac{3}{2} = -\frac{\sqrt{11}}{2}i$

    $x = \frac{3}{2} + \frac{\sqrt{11}}{2}i$ or $x = \frac{3}{2} - \frac{\sqrt{11}}{2}i$

    The solution set is $\left\{\frac{3}{2} \pm \frac{\sqrt{11}}{2}i\right\}$.

33. $y^2 - \frac{3}{2}y - 5 = 0$

$\qquad y^2 - \frac{3}{2}y = 5$ Add 5 to both sides.

$y^2 - \frac{3}{2}y + \frac{9}{16} = 5 + \frac{9}{16}$ Complete the square by adding $\left[\frac{1}{2}\left(-\frac{3}{2}\right)\right]^2$ to both sides.

$\qquad \left(y - \frac{3}{4}\right)^2 = \frac{89}{16}$ Factor on the left, and simplify on the right.

$\qquad y - \frac{3}{4} = \pm\sqrt{\frac{89}{16}}$ Solve.

$\qquad y - \frac{3}{4} = \pm\frac{\sqrt{89}}{4}$

$y - \frac{3}{4} = \frac{\sqrt{89}}{4}$ or $y - \frac{3}{4} = -\frac{\sqrt{89}}{4}$

$y = \frac{3}{4} + \frac{\sqrt{89}}{4}$ or $y = \frac{3}{4} - \frac{\sqrt{89}}{4}$

$y = \frac{3 + \sqrt{89}}{4}$ or $y = \frac{3 - \sqrt{89}}{4}$

The solution set is $\left\{\frac{3 \pm \sqrt{89}}{4}\right\}$.

C

37. $3x^2 + 7x + 2 = 0$

$\qquad x^2 + \frac{7}{3}x + \frac{2}{3} = 0$ Divide both sides by 3.

$\qquad x^2 + \frac{7}{3}x = -\frac{2}{3}$ Subtract $\frac{2}{3}$ from both sides.

$x^2 + \frac{7}{3}x + \frac{49}{36} = -\frac{2}{3} + \frac{49}{36}$ Complete the square by adding $\left[\frac{1}{2}\left(\frac{7}{3}\right)\right]^2$ to both sides.

$\qquad \left(x + \frac{7}{6}\right)^2 = \frac{25}{36}$

$\qquad x + \frac{7}{6} = \pm\frac{5}{6}$

$x + \frac{7}{6} = \frac{5}{6}$ or $x + \frac{7}{6} = -\frac{5}{6}$

$x = -\frac{7}{6} + \frac{5}{6}$ or $x = -\frac{7}{6} - \frac{5}{6}$

$x = -\frac{1}{3}$ or $x = -2$

The solution set is $\left\{-\frac{1}{3}, -2\right\}$.

41. $\quad 3x^2 + 2x = 6$

$\quad\quad x^2 + \frac{2}{3}x = 2$ Divide both sides by 3.

$x^2 + \frac{2}{3}x + \frac{1}{9} = 2 + \frac{1}{9}$ Add $\left[\frac{1}{2}\left(\frac{2}{3}\right)\right]^2$ to both sides.

$\left(x + \frac{1}{3}\right)^2 = \frac{19}{9}$

$x + \frac{1}{3} = \pm \frac{\sqrt{19}}{3}$ Solve.

$x + \frac{1}{3} = \frac{\sqrt{19}}{3}$ or $\quad x + \frac{1}{3} = -\frac{\sqrt{19}}{3}$

$x = -\frac{1}{3} + \frac{\sqrt{19}}{3}$ or $\quad x = -\frac{1}{3} - \frac{\sqrt{19}}{3}$

$x = \frac{-1 + \sqrt{19}}{3}$ or $\quad x = \frac{-1 - \sqrt{19}}{3}$

The solution set is $\left\{\frac{-1 \pm \sqrt{19}}{3}\right\}$.

45. $\quad\quad\quad 6x^2 = -3x - 1$

$\quad 6x^2 + 3x + 1 = 0$ Write the equation in standard form.

$\quad\quad 6x^2 + 3x = -1$ Subtract 1 from both sides.

$\quad\quad x^2 + \frac{1}{2}x = -\frac{1}{6}$ Divide both sides by 6.

$x^2 + \frac{1}{2}x + \frac{1}{16} = -\frac{1}{6} + \frac{1}{16}$ Add $\left[\frac{1}{2}\left(\frac{1}{2}\right)\right]^2$ to both sides.

$\left(x + \frac{1}{4}\right)^2 = -\frac{5}{48}$ Factor on the left, and simplify on the right.

$x + \frac{1}{4} = \pm\sqrt{-\frac{5}{48}}$ Solve.

$\quad\quad\quad = \pm\sqrt{-\frac{15}{144}}$ Rationalize the denominator.

$x + \frac{1}{4} = \pm \frac{\sqrt{15}}{12}i$

$x + \frac{1}{4} = \frac{\sqrt{15}}{12}i$ or $x + \frac{1}{4} = -\frac{\sqrt{15}}{12}i$

$x = -\frac{1}{4} + \frac{\sqrt{15}}{12}i$ or $\quad x = -\frac{1}{4} - \frac{\sqrt{15}}{12}i$ Write the complex roots in the form a + bi.

The solution set is $\left\{-\frac{1}{4} \pm \frac{\sqrt{15}}{12}i\right\}$.

49. $3x^2 - 4x + 5 = 0$

$x^2 - \frac{4}{3}x + \frac{5}{3} = 0$ Divide both sides by 3.

$x^2 - \frac{4}{3}x = -\frac{5}{3}$ Subtract $\frac{5}{3}$ from both sides.

$x^2 - \frac{4}{3}x + \frac{4}{9} = -\frac{5}{3} + \frac{4}{9}$ Add $\left[\frac{1}{2}\left(-\frac{4}{3}\right)\right]^2$ to both sides.

$\left(x - \frac{2}{3}\right)^2 = -\frac{11}{9}$ Factor on the left; simplify on the right.

$x - \frac{2}{3} = \pm\sqrt{-\frac{11}{9}}$ Solve.

$x - \frac{2}{3} = \pm\frac{\sqrt{11}}{3}i$

$x - \frac{2}{3} = \frac{\sqrt{11}}{3}i$ or $x - \frac{2}{3} = -\frac{\sqrt{11}}{3}i$

$x = \frac{2}{3} + \frac{\sqrt{11}}{3}i$ or $x = \frac{2}{3} - \frac{\sqrt{11}}{3}i$ Standard form for complex roots.

The solution set is $\left\{\frac{2}{3} \pm \frac{\sqrt{11}}{3}i\right\}$.

D

53. The number of diagonals D that a polygon of n sides has can be expressed in the formula $2D = n(n - 3)$. If a polygon has 104 diagonals, how many sides does it have?

$2D = n(n - 3)$ Formula.
$2(104) = n(n - 3)$ Substitute 104 for D.
$208 = n^2 - 3n$

$208 + \frac{9}{4} = n^2 - 3n + \frac{9}{4}$ Add $\left[\frac{1}{2}(-3)\right]^2$ to both sides.

$\frac{841}{4} = \left(n - \frac{3}{2}\right)^2$

$\pm\sqrt{\frac{841}{4}} = n - \frac{3}{2}$

$\pm\frac{29}{2} = n - \frac{3}{2}$

$\frac{3}{2} \pm \frac{29}{2} = n$

$n = \frac{3}{2} + \frac{29}{2}$ or $n = \frac{3}{2} - \frac{29}{2}$

$n = 16$ or $n = -13$ Reject the negative answer since the number of diagonals cannot be negative.

The polygon has 16 sides.

57. The sum of two numbers is 6, and their product is 10. What are the numbers?

If x represents one of the numbers, then 6 - x represents the other since their sum is 6.

$$x(6 - x) = 10$$
$$6x - x^2 = 10$$
The product equals 10.
$$x^2 - 6x = -10$$
Divide both sides by -1.
$$x^2 - 6x + 9 = -10 + 9$$
Add $\left[\frac{1}{2}(-6)\right]^2$ to both sides.
$$(x - 3)^2 = -1$$
Factor on the left, and simplify on the right.
$$x - 3 = \pm\sqrt{-1}$$
Solve.
$$x - 3 = \pm i$$
$$x = 3 \pm i$$

One of the numbers is 3 + i, and the other is 3 - i.

E Maintain Your Skills

Write using rational exponents:

61. $3\sqrt{x^2 + 4} = 3(x^2 + 4)^{1/2}$

Simplify. Assume all variables represent positive numbers:

65. $\sqrt{360x^6 y^9} = \sqrt{36 \cdot 10 \cdot x^6 \cdot y^8 \cdot y}$
    $= \sqrt{36 x^6 y^8} \sqrt{10y}$
    $= \sqrt{6^2 (x^3)^2 (y^4)^2} \sqrt{10y}$
    $= 6x^3 y^4 \sqrt{10y}$

EXERCISES 5.4 QUADRATIC EQUATIONS SOLVED BY THE QUADRATIC FORMULA

A

Solve using the quadratic formula:

1. $x^2 + 9x + 14 = 0$ The equation is in standard form.

    $a = 1, b = 9, c = 14$ Identify a, b, and c.

    $x = \dfrac{-b \pm \sqrt{b^2 - 4ac}}{2a}$ Quadratic formula.

    $x = \dfrac{-9 \pm \sqrt{(9)^2 - 4(1)(14)}}{2(1)}$ Substitute.

    $x = \dfrac{-9 \pm \sqrt{81 - 56}}{2}$ Simplify.

    $x = \dfrac{-9 \pm \sqrt{25}}{2}$

    $x = \dfrac{-9 \pm 5}{2}$

    $x = \dfrac{-9 + 5}{2}$ or $x = \dfrac{-9 - 5}{2}$

    $x = -2$ $x = -7$

    The solution set is $\{-7, -2\}$.

5. $2x^2 - 10x - 72 = 0$

    $a = 2, b = -10, c = -72$ Identify a, b, and c.

    $x = \dfrac{-b \pm \sqrt{b^2 - 4ac}}{2a}$ Quadratic formula.

    $x = \dfrac{10 \pm \sqrt{(-10)^2 - 4(2)(-72)}}{2(2)}$ Substitute.

    $x = \dfrac{10 \pm \sqrt{100 + 576}}{4}$

    $x = \dfrac{10 \pm \sqrt{676}}{4}$

    $x = \dfrac{10 \pm 26}{4}$

    $x = \dfrac{10 + 26}{4}$ or $x = \dfrac{10 - 26}{4}$

    $x = 9$ $x = -4$

    The solution set is $\{-4, 9\}$.

9. $x^2 - 20x = -96$ Standard form.
   $a = 1, b = -20, c = 96$

   $$x = \frac{-b \pm \sqrt{b^2 - 4ac}}{2a}$$

   $$x = \frac{-(-20) \pm \sqrt{(-20)^2 - 4(1)(96)}}{2(1)}$$

   $$x = \frac{20 \pm \sqrt{400 - 384}}{2}$$

   $$x = \frac{20 \pm \sqrt{16}}{2}$$

   $$x = \frac{20 \pm 4}{2}$$

   $$x = \frac{20 + 4}{2} \quad \text{or} \quad x = \frac{20 - 4}{2}$$

   $x = 12 \quad\quad\quad\quad x = 8$

   The solution set is $\{8, 12\}$.

13. $\quad\quad\quad 4x^2 = -8x - 3$
    $4x^2 + 8x + 3 = 0$ Standard form.
    $a = 4, b = 8, c = 3$

    $$x = \frac{-b \pm \sqrt{b^2 - 4ac}}{2a} \quad\quad \text{Quadratic formula.}$$

    $$x = \frac{-(8) \pm \sqrt{(8)^2 - 4(4)(3)}}{2(4)} \quad\quad \text{Substitute.}$$

    $$x = \frac{-8 \pm \sqrt{64 - 48}}{8} \quad\quad \text{Simplify.}$$

    $$x = \frac{-8 \pm \sqrt{16}}{8}$$

    $$x = \frac{-8 \pm 4}{8}$$

    $x = -\frac{1}{2} \quad\quad x = -\frac{3}{2}$

    The solution set is $\left\{-\frac{3}{2}, -\frac{1}{2}\right\}$.

B

17. $x^2 - 4x - 7 = 0$ Standard form.
$a = 1, b = -4, c = -7$

$$x = \frac{-b \pm \sqrt{b^2 - 4ac}}{2a}$$

$$x = \frac{-(-4) \pm \sqrt{(-4)^2 - 4(1)(-7)}}{2(1)}$$

$$x = \frac{4 \pm \sqrt{16 + 28}}{2}$$

$$= \frac{4 \pm \sqrt{44}}{2}$$

$$x = \frac{4 \pm 2\sqrt{11}}{2} = \frac{2(2 + \sqrt{11})}{2}$$ Divide out the common factor, 2.

$x = 2 \pm \sqrt{11}$

The solution set is $\{2 \pm \sqrt{11}\}$.

21. $3x^2 - 15x + 21 = 0$
$a = 3, b = -15, c = 21$

$$x = \frac{-b \pm \sqrt{b^2 - 4ac}}{2a}$$

$$x = \frac{15 \pm \sqrt{225 - 4(3)(21)}}{6}$$

$$x = \frac{15 \pm \sqrt{225 - 252}}{6} = \frac{15 \pm \sqrt{-27}}{6}$$

$$x = \frac{15 \pm 3i\sqrt{3}}{6}$$

$$x = \frac{3(5 \pm i\sqrt{3})}{3(2)}$$ Factor the numerator to reduce.

$$x = \frac{5 \pm i\sqrt{3}}{2}$$

or $\frac{5}{2} \pm \frac{\sqrt{3}}{2}i$ Standard form for complex roots: $a + bi$.

The solution set is $\left\{\frac{5}{2} \pm \frac{\sqrt{3}}{2}i\right\}$.

25. $-2x^2 + 3x + 7 = 0$
    $a = -2, \ b = 3, \ c = 7$ Identify a, b, and c.

$$x = \frac{-b \pm \sqrt{b^2 - 4ac}}{2a}$$

$$x = \frac{-3 \pm \sqrt{9 - 4(-2)(7)}}{2(-2)}$$

$$x = \frac{-3 \pm \sqrt{65}}{-4}$$

or $\dfrac{\cancel{-1}(3 \pm \sqrt{65})}{\cancel{-1}(4)}$ Factor out a -1 in the numerator and denominator.

The solution set is $\left\{\dfrac{3 \pm \sqrt{65}}{4}\right\}$.

29. $6x^2 - 19ax + 15a^2 = 0$ Standard form.
    $a = 6, \ b = -19a, \ c = 15a^2$ Identify a, b, and c.

$$x = \frac{-b \pm \sqrt{b^2 - 4ac}}{2a}$$ Quadratic formula.

$$x = \frac{-(-19a) \pm \sqrt{(-19a)^2 - 4(6)(15a^2)}}{2(6)}$$ Substitute.

$$x = \frac{19a \pm \sqrt{361a^2 - 360a^2}}{12}$$

$$x = \frac{19a \pm \sqrt{a^2}}{12}$$

$$x = \frac{19a \pm a}{12}$$

$x = \dfrac{19a + a}{12}$ or $x = \dfrac{19a - a}{12}$

$x = \dfrac{5a}{3}$ $x = \dfrac{3a}{2}$

The solution set is $\left\{\dfrac{3a}{2}, \dfrac{5a}{3}\right\}$.

C

33. $6x^2 = -19x - 10$ Write the equation in
    $6x^2 + 19x + 10 = 0$ standard form.
    $a = 6, b = 19, c = 10$ Identify a, b, and c.

    $x = \dfrac{-(19) \pm \sqrt{(19)^2 - 4(6)(10)}}{2(6)}$ Substitute into the formula from memory.

    $x = \dfrac{-19 \pm \sqrt{361 - 240}}{12}$ Simplify.

    $x = \dfrac{-19 \pm \sqrt{121}}{12}$

    $x = \dfrac{-19 \pm 11}{12}$

    $x = \dfrac{-19 + 11}{12}$ or $x = \dfrac{-19 - 11}{12}$

    $x = -\dfrac{2}{3}$ $x = -\dfrac{5}{2}$

    The solution set is $\left\{-\dfrac{5}{2}, -\dfrac{2}{3}\right\}$.

37. $-6x^2 = 13x - 5$
    $6x^2 + 13x - 5 = 0$ Standard form.
    $a = 6, b = 13, c = -5$

    $x = \dfrac{-(13) \pm \sqrt{(13)^2 - 4(6)(-5)}}{2(6)}$ Substitute into the formula from memory.

    $x = \dfrac{-13 \pm \sqrt{169 + 120}}{12}$

    $x = \dfrac{-13 \pm \sqrt{289}}{12}$

    $x = \dfrac{-13 \pm 17}{12}$

    $x = \dfrac{-13 + 17}{12}$ or $x = \dfrac{-13 - 17}{12}$

    $x = \dfrac{1}{3}$ $x = -\dfrac{5}{2}$

    The solution set is $\left\{-\dfrac{5}{2}, \dfrac{1}{3}\right\}$.

41. $\quad x^2 + 2x + 3 = 3x^2 + 7x - 3$
    $-2x^2 - 5x + 6 = 0$ Standard form.
    $2x^2 + 5x - 6 = 0$ Multiply both sides by -1.
    $a = 2, b = 5, c = -6$

    $x = \dfrac{-b \pm \sqrt{b^2 - 4ac}}{2a}$

    $x = \dfrac{-(5) \pm \sqrt{(5)^2 - 4(2)(-6)}}{2(2)}$ Substitute.

    $x = \dfrac{-5 \pm \sqrt{25 + 48}}{4}$ Simplify.

    $x = \dfrac{-5 \pm \sqrt{73}}{4}$

    The solution set is $\left\{\dfrac{-5 \pm \sqrt{73}}{4}\right\}$.

45. $ax^2 - 3abx + 2b^2 = 0$ Standard form.
    $a = a, b = -3ab, c = 2b^2$ Identify a, b, and c.

    $x = \dfrac{-b \pm \sqrt{b^2 - 4ac}}{2a}$ Quadratic formula.

    $x = \dfrac{-(-3ab) \pm \sqrt{(-3ab)^2 - 4(a)(2b^2)}}{2(a)}$

    $x = \dfrac{3ab \pm \sqrt{9a^2b^2 - 8ab^2}}{2a}$

    $x = \dfrac{3ab \pm \sqrt{b^2(9a^2 - 8a)}}{2a}$ Simplify the radical.

    $x = \dfrac{3ab \pm b\sqrt{9a^2 - 8a}}{2a}$

    The solution set is $\left\{\dfrac{3ab \pm b\sqrt{9a^2 - 8a}}{2a}\right\}$.

D

49. A piece of metal that is 16 in. wide is to be formed into a gutter of rectangular shape. Both sides are to be bent up an equal amount. What will be the depth of the gutter if the cross-sectional area is 32 in.²?

Sketch a diagram to show the cross-section with an equal amount bent up on each side.

Formula:
$A = \ell \cdot w$

Formula for the area of the rectangle.

Substiute:
$32 = (16 - 2x)x$

$\ell = 12 - 2x$, $w = x$, $A = 32$.

Solve:
$$32 = 16x - 2x^2$$
$$2x^2 - 16x + 32 = 0$$
$$x^2 - 8x + 16 = 0$$
$$a = 1, \ b = -8, \ c = 16$$

Standard form.
Divide both sides by 2.
Identify a, b, and c.

$$x = \frac{-(-8) \pm \sqrt{(-8)^2 - 4(1)(16)}}{2(1)}$$

Substitute into the formula by memory.

$$x = \frac{8 \pm \sqrt{64 - 64}}{2} = \frac{8 \pm 0}{2}$$

$$x = 4$$

A single root, multiplicity two.

The gutter is 4 in. deep.

E Maintain Your Skills

Simplify using only positive exponents:

53. $\left[\dfrac{a^{5/8}}{a^{3/4}}\right]^{-1/2} = \dfrac{(a^{5/8})^{-1/2}}{(a^{3/4})^{-1/2}}$

$= \dfrac{a^{-5/16}}{a^{-3/8}}$

$= \dfrac{1}{a^{5/16}} \cdot \dfrac{a^{3/8}}{1}$

$= \dfrac{a^{3/8}}{a^{5/16}} = a^{3/8 - 5/16} = a^{1/16}$

57. $(a^{1/2} - 5)(a^{1/2} + 3)$

$= (a^{1/2})(a^{1/2}) + (a^{1/2})(3) + (-5)(a^{1/2}) + (-5)(3)$

$= a^{1/2 + 1/2} + 3a^{1/2} - 5a^{1/2} - 15$

$= a - 2a^{1/2} - 15$

Simplify:

61. $\dfrac{1 + \dfrac{1}{a-b}}{3 - \dfrac{1}{a+b}} = \dfrac{(a+b)(a-b)\left[1 + \dfrac{1}{a-b}\right]}{(a+b)(a-b)\left[3 - \dfrac{1}{a+b}\right]}$

$= \dfrac{(a+b)(a-b) + (a+b)}{3(a+b)(a-b) - (a-b)}$

$= \dfrac{a^2 - b^2 + a + b}{3a^2 - 3b^2 - a + b}$

EXERCISES 5.5   PROPERTIES OF QUADRATIC EQUATIONS

A

Find the value of the discriminant and describe the roots of each of the following:

1. $x^2 + 10x + 16 = 0$
   $a = 1, b = 10, c = 16$ — Identify a, b, and c.
   $b^2 - 4ac = (10)^2 - 4(1)(16)$ — Substitute into the formula
   $= 100 - 64$ — for the discriminant.
   $= 36$ — The discriminant is positive, so the equation has two unequal real roots.

   The discriminant is 36, and the equation has two unequal real roots.

5. $x^2 + 16x + 64 = 0$
   $a = 1, b = 16, c = 64$ — Identify a, b, and c.
   $b^2 - 4ac = (16)^2 - 4(1)(64)$ — Substitute into the formula for the discriminant.
   $= 256 - 256$
   $= 0$

   The value of the discriminant is zero, and the equation has two equal real roots.

9. $x^2 = -x + 11$
   $x^2 + x - 11 = 0$ — Standard form.
   $a = 1, b = 1, c = -11$ — Identify a, b, and c.
   $b^2 - 4ac = (1)^2 - 4(1)(-11)$ — Substitute into the formula for the discriminant.
   $= 1 + 44$
   $= 45$

   The discriminant is 45, and the equation has two unequal real roots.

Write a quadratic equation in standard form that has the following roots:

13. $x = 7$ or $x = -5$
    $x - 7 = 0$ or $x + 5 = 0$ — Rewrite each equation so the right side is zero.
    $(x - 7)(x + 5) = 0$ — Zero-product law.
    $x^2 - 2x - 35 = 0$ — Write in standard form.

B

Find the value of the discriminant and describe the roots of each of the following:

17. $3x^2 - 8x + 11 = 0$
      $a = 3, b = -8, c = 11$ Identify a, b, and c.
      $b^2 - 4ac = (-8)^2 - 4(3)(11)$ Substitute.
              $= 64 - 132$
              $= -68$

The discriminant is negative, and the equation has two conjugate complex roots.

21. $7x^2 - 2x = -1$
      $7x^2 - 2x + 1 = 0$ Standard form.
      $a = 7, b = -2, c = 1$
      $b^2 - 4ac = (-2)^2 - 4(7)(1)$ Substitute.
              $= 4 - 28$
              $= -24$

The discriminant is negative, so the equation has two conjugate complex roots.

25. $7x^2 - 2x = x + 14$
      $7x^2 - 3x - 14 = 0$ Standard form.
      $a = 7, b = -3, c = -14$ Identify a, b, and c.
      $b^2 - 4ac = (-3)^2 - 4(7)(-14)$ Substitute into the formula
                                          for the discriminant.
              $= 9 + 392$
              $= 401$

The discriminant is 401, and the equation has two unequal real roots.

Write a quadratic equation in standard form that has the following roots:

29. $x = 1 + \sqrt{5}$ or $x = 1 - \sqrt{5}$
      $x - 1 - \sqrt{5} = 0$ or $x - 1 + \sqrt{5} = 0$ Rewrite each equation so that the right side is zero.
            $(x - 1 - \sqrt{5})(x - 1 + \sqrt{5}) = 0$ Zero-product property.
            $[(x - 1) - \sqrt{5}][(x - 1) + \sqrt{5}] = 0$ Write factors on left side as conjugates.
                  $[(x - 1)^2 - (\sqrt{5})^2] = 0$ Product of conjugates.
                      $x^2 - 2x + 1 - 5 = 0$ Multiply.
                      $x^2 - 2x - 4 = 0$ Standard form.

C

Find the value of the discriminant and describe the roots of each of the following:

33. $\quad 7x^2 + 4x = 3x - 10$
$7x^2 + x + 10 = 0$ Rewrite in standard form.
$a = 7, b = 1, c = 10$ Identify a, b, and c.
$b^2 - 4ac = (1) - 4(7)(10)$ Substitute into the formula for the discriminant.
$\qquad = 1 - 280$
$\qquad = -279$

The discriminant is -279 so the equation has two conjugate complex roots.

37. $\quad 2x^2 - x + 2 = -2x^2 + 11x - 7$
$4x^2 - 12x + 9 = 0$ Rewrite in standard form.
$a = 4, b = -12, c = 9$ Identify a, b, and c.
$b^2 - 4ac = (-12)^2 - 4(4)(9)$ Substitute.
$\qquad = 144 - 144$
$\qquad = 0$

The discriminant is zero, so the equation has two equal real roots.

41. $\quad (3x - 1)^2 - (x + 5)^2 = 6$
$9x^2 - 6x + 1 - (x^2 + 10x + 25) = 6$ Multiply and simplify.
$9x^2 - 6x + 1 - x^2 - 10x - 25 = 6$
$8x^2 - 16x - 24 = 6$
$8x^2 - 16x - 30 = 0$ Rewrite in standard form.
$a = 8, b = -16, c = -30$ Identify a, b, and c.
$b^2 - 4ac = (-16)^2 - 4(8)(-30)$ Substitute.
$\qquad = 256 + 960$
$\qquad = 1216$

The discriminant is 1216, so the equation has two unequal real roots.

Write an equation in standard form that has the following roots:

45. $x = \dfrac{2 \pm \sqrt{5}}{3}$

$x = \dfrac{2 + \sqrt{5}}{3}$ or $x = \dfrac{2 - \sqrt{5}}{3}$

$3x = 2 + \sqrt{5}$ or $3x = 2 - \sqrt{5}$	Multiply both sides of each equation by 3 to clear the fraction.
$3x - 2 - \sqrt{5} = 0$ or $3x - 2 + \sqrt{5} = 0$	Write each equation with the right side equal to zero.
$(3x - 2 - \sqrt{5})(3 - 2 + \sqrt{5}) = 0$	Zero-product property.
$[(3x - 2) - \sqrt{5}][(3x - 2) + \sqrt{5}] = 0$	Write the left side using conjugate factors.
$(3x - 2)^2 - (\sqrt{5})^2 = 0$	Product of conjugates.
$9x^2 - 12x + 4 - 5 = 0$	Multiply.
$9x^2 - 12x - 1 = 0$	Standard form.

49. $x = -3 \pm 2i\sqrt{5}$

$x = -3 + 2i\sqrt{5}$ or $\qquad\qquad x = -3 - 2i\sqrt{5}$

$x + 3 - 2i\sqrt{5} = 0$ or $x + 3 + 2i\sqrt{5} = 0$

$[(x + 3) - 2i\sqrt{5}][(x + 3) + 2i\sqrt{5}] = 0$

$(x + 3)^2 - (2i\sqrt{5})^2 = 0$

$x^2 + 6x + 9 - 4i^2(5) = 0$

$x^2 + 6x + 29 = 0 \qquad\qquad i^2 = -1$

D

53. For what value of k will $x^2 + kx + 21 = 0$ have two equal real solutions?

$b^2 - 4ac = 0$	The equation will have two equal real solutions if $b^2 - 4ac$ is equal to zero.
$a = 1, b = k, c = 21$	
$(k^2) - 4(1)(21) = 0$	Substitute.
$k^2 - 84 = 0$	Multiply.
$k^2 = 84$	Solve for k
$k = \pm\sqrt{84}$	
$k = \pm 2\sqrt{21}$	

When $k = \pm 2\sqrt{21}$, the equation has two equal real solutions.

E Maintain Your Skills

Write in radical form:

57. $(x^2 - x + 1)^{1/4} = \sqrt[4]{x^2 - x + 1}$ The denominator, 4, of the fractional exponent represents the fourth root.

Simplify. Assume that variables represent real numbers:

61. $9\sqrt{27x^6y^8} = 9\sqrt{9 \cdot 3 \cdot x^6 \cdot y^8}$ Both variables have exponents that are even, so no absolute value signs are necessary.
$= 27x^3y^4\sqrt{3}$

Combine. Assume that variables represent positive numbers:

65. $\sqrt{13y^4} - y^2\sqrt{52} + \sqrt{325y^4}$
$= y^2\sqrt{13} - y^2\sqrt{4 \cdot 13} + \sqrt{25 \cdot 13 \cdot y^4}$
$= y^2\sqrt{13} - 2y^2\sqrt{13} + 5y^2\sqrt{13}$
$= (y^2 - 2y^2 + 5y^2)\sqrt{13}$
$= 4y^2\sqrt{13}$

EXERCISES 5.6 EQUATIONS INVOLVING ABSOLUTE VALUE

A

Solve:

1. $|x| = 3$
    $x = \pm 3$

    3 and -3 are the two numbers that are three units from zero.

    The solution set is $\{\pm 3\}$.

5. $|x - 4| = 5$
    $x - 4 = 5$ or $x - 4 = -5$
    $x = 9$ $x = -1$

    If $|x| = c$, $c \geq 0$, then $x = c$ or $x = -c$.

    The solution set is $\{-1, 9\}$.

9. $|x + 5| = 0$
    $x + 5 = 0$
    $x = -5$

    The solution set is $\{-5\}$.

13. $|x - 9| = 20$
    $x - 9 = 20$ or $x - 9 = -20$
    $x = 29$ $x = -11$

    The solution set is $\{-11, 29\}$.

B

17. $|x + 3| - 8 = 5$
    $|x + 3| = 13$ Isolate the absolute value on
                                                one side of the equation.
    $x + 3 = 13$ or $x + 3 = -13$
    $x = 10$ $x = -16$

    The solution set is $\{-16, 10\}$.

21. $|3x + 6| = 9$
    $3x + 6 = 9$ or $3x + 6 = -9$
    $3x = 3$ $3x = -15$
    $x = 1$ $x = -5$

    The solution set is $\{-5, 1\}$.

25. $-2|2x - 14| = 40$

    $|2x - 14| = -20$ Multiply both sides by the
                                                reciprocal of -2.
    No solution.   An absolute value cannot be
                                                negative.

    The solution set is empty, $\emptyset$.

29. $6 - |2x + 7| = -4$

    $|2x + 7| = 10$ Isolate the absolute value on
                                                one side of the equation.
    $2x + 7 = 10$ or $2x + 7 = -10$
    $2x = 3$ $2x = -17$
    $x = \frac{3}{2}$ $x = \frac{-17}{2}$

    The solution set is $\left\{-\frac{17}{2}, \frac{3}{2}\right\}$.

C

33. $|x| = 2x - 3$

$2x - 3 \geq 0$
$2x \geq 3$
$x \geq \frac{3}{2}$

Restrict the variable so that $|x|$ will not be negative, so $2x - 3 \geq 0$.

$x = 2x - 3$ or $x = -(2x - 3)$
$-x = -3$ $x = -2x + 3$
$x = 3$ $3x = 3$
                  $x = 1$

$x = cx + d$ or $x = -(cx + d)$.

Reject $x = 1$ since it is a condition that $x \geq \frac{3}{2}$.

Check:
$|x| = 2x - 3$

$|3| = 2(3) - 3$
$3 = 6 - 3$
$3 = 3$

Substitute $x = 3$.

The statement is true.

The solution set is {3}.

37. $-3|x - 5| = 9 - 3x$

$-3|x - 5| = -3(x - 3)$ Factor on the right.

$|x - 5| = x - 3$ Divide both sides by the common factor, -3.

$x - 5 = x - 3$ or $x - 5 = -(x - 3)$ If $|x| = c$, then $x = c$
$-5 = -3$ $x - 5 = -x + 3$ or $x = -c$.
Contradiction $2x = 8$
                   $x = 4$

The solution set is {4}.

41. $|2x + 3| = 6 - x$

$6 - x \geq 0$
$-x \geq -6$
$x \leq 6$

Restrict the variable so that $|2x + 3|$ cannot be negative, so $6 - x \geq 0$.

$2x + 3 = 6 - x$ or $2x + 3 = -(6 - x)$
$3x = 3$ $2x + 3 = -6 + x$
$x = 1$ $x = -9$

Check:
$|2x + 3| = 6 - x$

$|2(1) + 3| = 6 - 1$
$|5| = 5$
$5 = 5$

Substitute $x = 1$.

$|2(-9) + 3| = 6 - (-9)$
$|-18 + 3| = 6 + 9$
$|-15| = 15$
$15 = 15$

Substitute $x = -9$.

The solution set is {−9, 1}.

45. $|7x - 2| - 13 = 3x$
    $|7x - 2| = 3x + 13$ Isolate the absolute value on one side of the equation.

   $7x - 2 = 3x + 13$ or $7x - 2 = -(3x + 13)$
   $4x = 15$ $7x - 2 = -3x - 13$
   $x = \frac{15}{4}$ $10x = -11$
                             $x = \frac{-11}{10}$

   The solution set is $\left\{-\frac{11}{10}, \frac{15}{4}\right\}$.

D

49. The weekly cost (C) of producing crowbars at the Old Crow Manufacturing Company is given by C = 10000 + 2.5x where x is the number of crowbars produced. Find the maximum and minimum number of crowbars if the weekly costs are held at $25,000 plus or minus $3000.

   Simpler word form:
   $25000 minus weekly cost = ±3000.

   In absolute value form:
   $|\$25000 - \text{weekly cost}| = 3000$

   Select a variable:
   Let x represent the number of crobars.

   Translate to algebra:
   $|25000 - (10000 + 2.5x)| = 3000$ The formula for cost is given in the problem: 10000 + 2.5x.

   Solve:
   $|15000 - 2.5x| = 3000$
   $15000 - 2.5x = 3000$ or $15000 - 2.5x = -3000$
   $-2.5x = -12000$ $-2.5x = -18000$
   $x = 4800$ $x = 7200$

   The maximum number of crowbars is 7200, and the minimum number is 4800.

E Maintain Your Skills

53. $(6\sqrt{14})(3\sqrt{21}) = 18\sqrt{294}$
                          $= 18\sqrt{49}\sqrt{6}$
                          $= 18(7)\sqrt{6} = 126\sqrt{6}$

57. $\sqrt{6}(\sqrt{30} - \sqrt{54}) = \sqrt{180} - \sqrt{324}$
$= \sqrt{36}\sqrt{5} - 18$
$= 6\sqrt{5} - 18$

61. Find the perimeter of a hexagonal flower bed that measures $\sqrt{75}$ ft, $\sqrt{300}$ ft, $\sqrt{108}$ ft, $\sqrt{48}$ ft, $\sqrt{243}$ ft, and $\sqrt{147}$ ft on a side. Give the answer
   a. in simplest radical form.
   b. to the nearest tenth of a foot.

   Formula:
   P = sum of the sides.

   Substitute:
   $P = \sqrt{75} + \sqrt{300} + \sqrt{108} + \sqrt{48} + \sqrt{243} + \sqrt{147}$

   Solve:
   $P = \sqrt{25}\sqrt{3} + \sqrt{100}\sqrt{3} + \sqrt{36}\sqrt{3} + \sqrt{16}\sqrt{3} + \sqrt{81}\sqrt{3} + \sqrt{49}\sqrt{3}$
   $P = \sqrt{3}(5 + 10 + 6 + 4 + 9 + 7)$
   $P = 41\sqrt{3} \approx 71.0$ Use a calculator to approximate the value of $41\sqrt{3}$.

   The perimeter is
   a. $41\sqrt{3}$ ft.
   b. 71.0 ft.

EXERCISES 5.7   RADICAL EQUATIONS

A

1. $\sqrt{a} = 11$ Square both sides to eliminate the radical.

   $(\sqrt{a})^2 = (11)^2$
   $a = 121$ Simplify.

   Check:
   $\sqrt{121} = 11$ Substitute a = 121.
   $11 = 11$ True.

   The solution set is {121}.

5. $\sqrt{x} + 4 = 5$

    $\sqrt{x} = 1$ Isolate the radical on the left side.

    $(\sqrt{x})^2 = (1)^2$ Square both sides to eliminate the radical.

    $x = 1$ Simplify.

    Check:

    $\sqrt{1} + 4 = 5$ Substitute $x = 1$.
    $1 + 4 = 5$
    $5 = 5$ True.

    The solution set is {1}.

9. $\sqrt{2x + 1} = 3$
    $2x + 1 = 9$ Square both sides.
    $2x = 8$ Simplify.
    $x = 4$

    The solution set is {4}.

13. $5 - \sqrt{2x} = -11$

    $-\sqrt{2x} = -16$ Isolate the radical

    $(-\sqrt{2x})^2 = (-16)^2$ Square both sides.

    $2x = 256$
    $x = 128$

    Check:

    $5 - \sqrt{2 \cdot 128} = -11$ Replace x with 128.
    $5 - \sqrt{256} = -11$
    $5 - 16 = -11$
    $-11 = -11$ True.

    The solution set is {128}.

B

17.
$$\sqrt{3 - 4x} = 2x$$
$$(\sqrt{3 - 4x})^2 = (2x)^2 \quad \text{Square both sides.}$$
$$3 - 4x = 4x^2$$
$$4x^2 + 4x - 3 = 0$$
$$(2x + 3)(2x - 1) = 0 \quad \text{Solve by factoring.}$$
$$x = -\frac{3}{2} \quad \text{or} \quad x = \frac{1}{2}$$

Check:

$$\sqrt{3 - 4\left[-\frac{3}{2}\right]} = 2\left[-\frac{3}{2}\right] \quad \text{Substitute } x = -\frac{3}{2}.$$
$$\sqrt{3 + 6} = -3$$
$$\sqrt{9} = -3 \quad \text{Reject since, by definition,}$$
$$\sqrt{9} \text{ must be non-negative.}$$

$$\sqrt{3 - 4\left[\frac{1}{2}\right]} = 2\left[\frac{1}{2}\right] \quad \text{Substitute } x = \frac{1}{2}.$$
$$\sqrt{3 - 2} = 1$$
$$\sqrt{1} = 1$$
$$1 = 1 \quad \text{True.}$$

The solution set is $\left\{\frac{1}{2}\right\}$.

21.
$$\sqrt{x + 2} = x + 2$$
$$x + 2 = x^2 + 4x + 4 \quad \text{Square both sides.}$$
$$0 = x^2 + 3x + 2$$
$$0 = (x + 2)(x + 1) \quad \text{Factor.}$$
$$x + 2 = 0 \text{ or } x + 1 = 0 \quad \text{Zero-product property.}$$
$$x = -2 \text{ or } \quad x = -1$$

Check:

$$\sqrt{x + 2} = x + 2 \quad \text{Substitute } -2 \text{ for } x.$$
$$\sqrt{-2 + 2} = -2 + 2$$
$$0 = 0 \quad \text{True.}$$
$$\sqrt{x + 2} = x + 2 \quad \text{Substitute } -1 \text{ for } x.$$
$$\sqrt{-1 + 2} = -1 + 2$$
$$1 = 1 \quad \text{True. } \sqrt{1} = 1.$$

The solution set is $\{-2, -1\}$.

164

25. $\sqrt{12x + 13} = 3x - 2$
    $12x + 13 = 9x^2 - 12x + 4$ Square both sides.
    $0 = 9x^2 - 24x - 9$ Standard form.
    $0 = 3(3x + 1)(x - 3)$ Factor.
    $3x + 1 = 0$ or $x - 3 = 0$
    $x = -\frac{1}{3}$ or $x = 3$

Check:

$\sqrt{12\left[-\frac{1}{3}\right] + 13} = 3\left[-\frac{1}{3}\right] - 2$ Substitute $-\frac{1}{3}$ for x.
$\sqrt{9} = -3$
$3 = -3$ False.

$\sqrt{12(3) + 13} = 3(3) - 2$
$\sqrt{49} = 9 - 2$
$7 = 7$ True.

The solution set is {3}.

29. $\sqrt{2x + 5} + 5 = x$
    $\sqrt{2x + 5} = x - 5$ Isolate the radical.
    $(\sqrt{2x + 5})^2 = (x - 5)^2$ Square both sides.
    $2x + 5 = x^2 - 10x + 25$ Simplify.
    $x^2 - 12x + 20 = 0$
    $(x - 10)(x + 2) = 0$ Solve by factoring.
    $x = 10$ or $x = 2$

Check:

$\sqrt{2 \cdot 10 + 5} + 5 = 10$ Substitute $x = 10$.
$\sqrt{25} + 5 = 10$
$5 + 5 = 10$
$10 = 10$ True.

$\sqrt{2 \cdot 2 + 5} + 5 = 2$ Substitute $x = 2$.
$\sqrt{9} + 5 = 2$
$8 = 2$ Contradiction, therefore, reject $x = 2$.

The solution set is {10}.

C

33. $\sqrt{x} - \sqrt{x+9} = -1$ The equation contains two radicals.

$\sqrt{x} = \sqrt{x+9} - 1$ Isolate $\sqrt{x}$ on the left.
$(\sqrt{x})^2 = (\sqrt{x+9} - 1)^2$ Square both sides.
$x = x + 9 - 2\sqrt{x+9} + 1$
$x = x + 10 - 2\sqrt{x+9}$ Simplify.
$-10 = -2\sqrt{x+9}$
$5 = \sqrt{x+9}$ Divide both sides by -2 to isolate the second radical.
$(5)^2 = (\sqrt{x+9})^2$ Square both sides.
$25 = x + 9$ Solve for x.
$16 = x$

Check:
$\sqrt{16} - \sqrt{16+9} = -1$ Substitute $x = 16$.
$4 - \sqrt{25} = -1$
$4 - 5 = -1$
$-1 = -1$ True.

The solution set is $\{16\}$.

37. $\sqrt{x+1} + \sqrt{x} = -7$ The equation contains two radicals.

$\sqrt{x+1} = -\sqrt{x} - 7$ Isolate $\sqrt{x+1}$.
$(\sqrt{x+1})^2 = (-\sqrt{x} - 7)^2$ Square both sides.
$x + 1 = x + 14\sqrt{x} + 49$
$-48 = 14\sqrt{x}$ Simplify.
$-\dfrac{24}{7} = \sqrt{x}$ Divide by 14.

The equation has no roots since, by definition $\sqrt{x}$ must be non-negative.

The solution set is empty, $\{\ \}$, or $\emptyset$.

41. $\quad\sqrt{2y} + 3 = \sqrt{y - 7}$ The equation contains two radicals. The one on the right is isolated.

$\quad(\sqrt{2y} + 3)^2 = (\sqrt{y - 7})^2$ Square both sides.

$2y + 6\sqrt{2y} + 9 = y - 7$

$\quad\quad\quad 6\sqrt{2y} = -y - 16$ Simplify.

$\quad\quad\quad\sqrt{2y} = \dfrac{-y - 16}{6}$ Divide by 6 to isolate the second radical.

$\quad\quad(\sqrt{2y})^2 = \left[\dfrac{-y - 16}{6}\right]^2$ Square both sides.

$\quad\quad\quad 2y = \dfrac{y^2 + 32y + 256}{36}$

$\quad\quad\quad 72y = y^2 + 32y + 256$ Multiply both sides by 36.

$y^2 - 40y + 256 = 0$ Standard form.

$(y - 32)(y - 8) = 0$ Factor to solve for y.

$y - 32 = 0$ or $y - 8 = 0$

$\quad y = 32$ or $\quad y = 8$

Check:

$\sqrt{2 \cdot 32} + 3 = \sqrt{32 - 7}$ Substitute $y = 32$.

$\quad 8 + 3 = 5$

$\quad\quad 11 = 5$ Contradiction.

$\sqrt{2 \cdot 8} + 3 = \sqrt{8 - 7}$ Substitute $y = 8$.

$\quad 4 + 3 = 1$

$\quad\quad 7 = 1$ Contradiction.

Therefore, the solution set is { } or ∅.

45. $\quad\sqrt{5w + 9} = 3 + \sqrt{w}$

$\quad(\sqrt{5w + 9})^2 = (3 + \sqrt{w})^2$ Square both sides, since the radical on the left is isolated.

$5w + 9 = 9 + 6\sqrt{w} + w$

$\quad\quad 4w = 6\sqrt{w}$ Simplify.

$\quad\quad \dfrac{2}{3}w = \sqrt{w}$ Divide both sides by 6 to isolate the second radical.

$\quad\left[\dfrac{2}{3}w\right]^2 = (\sqrt{w})^2$ Square both sides.

$\quad\quad \dfrac{4}{9}w^2 = w$

$\quad\quad \dfrac{4}{9}w^2 - w = 0$ Write the equation in standard form.

$w\left[\dfrac{4}{9}w - 1\right] = 0$ Solve for w by factoring.

$w = 0$ or $\dfrac{4}{9}w - 1 = 0$

So $w = 0$ or $w = \dfrac{9}{4}$

Check:

$\sqrt{5 \cdot 0 + 9} = 3 + \sqrt{0}$ Substitute $w = 0$.
$\sqrt{9} = 3$
$3 = 3$ True.

$\sqrt{5\left[\frac{9}{4}\right] + 9} = 3 + \sqrt{\frac{9}{4}}$ Substitute $w = \frac{9}{4}$.

$\sqrt{\frac{45}{4} + \frac{36}{4}} = 3 + \frac{3}{2}$

$\sqrt{\frac{81}{4}} = \frac{6}{2} + \frac{3}{2}$

$\frac{9}{2} = \frac{9}{2}$ True.

The solution set is $\left\{0, \frac{9}{4}\right\}$.

D

49. Use this formula: $G = 26.8d^2\sqrt{p}$ to find the pressure of water from a fire hydrant outlet that is 4.5 inches in diameter and that discharges at 900 gallons per minute (round to nearest tenth).

Formula:
$G = 26.8d^2\sqrt{p}$

Substitute:
$900 = 26.8 (4.5)^2 \sqrt{p}$ To find the pressure, p, substitute $G = 900$ and $d = 4.5$.

Solve:
$900 = 542.7\sqrt{p}$
$\frac{900}{542.7} = \sqrt{p}$ Isolate the radical.
$1.66 \approx \sqrt{p}$
$2.75 \approx p$ Square both sides.

Answer:
The pressure to the nearest tenth is 2.8 psi (pressure in pounds per square inch).

53. The interest rate (r) (compounded annually) needed to have P dollars grow to A dollars at the end of two years is given by $r = \sqrt{\frac{A}{P}} - 1$. Find the value of A if P = 2000 and r = 0.1.

Formula:
$r = \sqrt{\frac{A}{P}} - 1$

Substitute:

$0.1 = \sqrt{\frac{A}{2000}} - 1$	r = 0.1 and P = 2000.
$1.1 = \sqrt{\frac{A}{2000}}$	Isolate the radical.
$1.21 = \frac{A}{2000}$	Square both sides.
$2420 = A$	Solve for A.

Answer:
The value of A is $2420.

E Maintain Your Skills

Combine and simplify:

57. $\sqrt{405w^3} + \sqrt{720w^3}$

$= \sqrt{81w^2}\sqrt{5w} + \sqrt{144w^2}\sqrt{5w}$

$= 9w\sqrt{5w} + 12w\sqrt{5w}$

$= 21w\sqrt{5w}$

Combine:

61. $4 + \frac{2}{x-3} - \frac{2x}{x+5} + \frac{5}{x^2+2x-15}$

$= \frac{4(x-3)(x+5) + 2(x+5) - 2x(x-3) + 5}{(x-3)(x+5)}$

$= \frac{4x^2 + 8x - 60 + 2x + 10 - 2x^2 + 6x + 5}{(x-3)(x+5)}$

$= \frac{2x^2 + 16x - 45}{(x-3)(x+5)}$

EXERCISES 5.8   FRACTIONAL EQUATIONS EQUIVALENT TO QUADRATIC EQUATIONS

A

Solve:

1. $\dfrac{x + 1}{2x + 1} = 5$

   $x \neq -\dfrac{1}{2}$ Restrict the variable to avoid division by zero. Multiply by the LCD.
   $x + 1 = 5(2x + 1)$
   $x + 1 = 10x + 5$
   $-9x = 4$
   $x = -\dfrac{4}{9}$

   The solution set is $\left\{-\dfrac{4}{9}\right\}$.

5. $\dfrac{x^2 + 1}{x + 3} = \dfrac{11}{x + 3}$

   $x \neq -3$ Restrict the variable. Multiply both sides by the LCM, $x + 3$.
   $x^2 + 1 = 11$ Solve for x.
   $x^2 = 10$
   $x = \pm\sqrt{10}$ Neither of these values is restricted.

   The solution set is $\{\pm\sqrt{10}\}$.

9. $\dfrac{2}{x} = x - 1$

   $x \neq 0$ Restrict the variable. Multiply both sides by the LCM, x.
   $2 = x(x - 1)$
   $2 = x^2 - x$
   $x^2 - x - 2 = 0$ Write in standard form. Factor to solve.
   $(x + 1)(x - 2) = 0$
   $x = -1$ or $x = 2$ Neither of these values is restricted.

   The solution set is $\{-1, 2\}$.

13. $\dfrac{12}{y + 6} = y - 5$

    $y \neq -6$ Restrict the variable. Multiply both sides by the LCM, $y + 6$.
    $12 = (y + 6)(y - 5)$
    $12 = y^2 + y - 30$
    $y^2 + y - 42 = 0$ Write in standard form. Factor and solve.
    $(y + 7)(y - 6) = 0$
    $y = -7$ or $y = 6$ Neither value is restricted.

    The solution set is $\{-7, 6\}$.

B

17. $\dfrac{3}{x + 4} + x = \dfrac{15}{x + 4}$

$\qquad\qquad x \neq -4$ — Restrict the variable.

$\qquad 3 + x(x + 4) = 15$ — Multiply both sides by the LCM, $x + 4$.
$\qquad 3 + x^2 + 4x = 15$
$\qquad x^2 + 4x - 12 = 0$ — Write in standard form.
$\qquad (x + 6)(x - 2) = 0$ — Factor and solve for x.
$\qquad x = -6 \text{ or } x = 2$ — Neither value is restricted.

The solution set is $\{-6, 2\}$.

21. $\dfrac{2}{x - 1} + \dfrac{x}{x - 1} = 2x$

$\qquad\qquad x \neq 1$ — Restrict the variable.

$\qquad 2 + x = 2x(x - 1)$ — Multiply both sides by the LCM, $x - 1$.
$\qquad 2 + x = 2x^2 - 2x$
$\qquad 2x^2 - 3x - 2 = 0$ — Write in standard form.
$\qquad (2x + 1)(x - 2) = 0$ — Factor and solve for x.
$\qquad x = -\dfrac{1}{2} \text{ or } x = 2$ — Neither value is restricted.

The solution set is $\left\{-\dfrac{1}{2}, 2\right\}$.

25. $\dfrac{3}{x - 3} - \dfrac{x^2 + 7}{x^2 - x - 6} = \dfrac{-1}{x + 2}$

$\qquad \dfrac{3}{x - 3} - \dfrac{x^2 + 7}{(x - 3)(x + 2)} = \dfrac{-1}{x + 2}$ — Factor the second denominator.

$\qquad x \neq 3 \text{ and } x \neq -2$ — Restrict the variables.

$\qquad 3(x + 2) - (x^2 + 7) = -1(x - 3)$ — Multiply both sides by the LCM.

$\qquad 3x + 6 - x^2 - 7 = -x + 3$
$\qquad x^2 - 4x + 4 = 0$ — Write in standard form. Factor.
$\qquad (x - 2)^2 = 0$
$\qquad x = 2 \text{ or } x = 2$ — A double root, and unrestricted value.

The solution set is $\{2\}$.

171

29. $x + 7 = \dfrac{1}{x} + \dfrac{x^2 - 8}{x - 3}$

$\qquad x \neq 0, 3$ Restrict the variable.

$\qquad x(x - 3)(x + 7) = 1(x - 3) + x(x^2 - 8)$ Multiply by the LCM, $x(x - 3)$.

$\qquad x^3 + 4x^2 - 21x = x - 3 + x^3 - 8x$ Simplify.
$\qquad 4x^2 - 14x + 3 = 0$ Write in standard form.
$\qquad a = 4, \; b = -14, \; c = 3$ Solve by using the quadratic formula.

$\qquad x = \dfrac{14 \pm \sqrt{196 - 4(4)(3)}}{2(4)}$

$\qquad x = \dfrac{14 \pm \sqrt{148}}{8}$

$\qquad x = \dfrac{14 \pm \sqrt{4}\sqrt{37}}{8}$

$\qquad x = \dfrac{14 \pm 2\sqrt{37}}{8} = \dfrac{2(7 \pm \sqrt{37})}{2(4)}$ Divide out the common factor, 2.

$\qquad x = \dfrac{7 \pm \sqrt{37}}{4}$ Neither is a restricted value.

The solution set is $\left\{\dfrac{7 \pm \sqrt{37}}{4}\right\}$.

C

33. $\dfrac{x - 1}{x^2 + x - 6} + \dfrac{x - 2}{x^2 + 4x + 3} = \dfrac{x + 1}{x^2 - x - 2}$

$\dfrac{x - 1}{(x + 3)(x - 2)} + \dfrac{x - 2}{(x + 3)(x + 1)} = \dfrac{x + 1}{(x - 2)(x + 1)}$ Factor the denominators.

$x \neq -3, -1, 2.$ Restrict the variable.

$(x - 1)(x + 1) + (x - 2)(x - 2) = (x + 1)(x + 3)$ Multiply by the LCM $(x+3)(x-2)(x+1)$.

$x^2 - 1 + x^2 - 4x + 4 = x^2 + 4x + 3$
$\qquad\qquad x^2 - 8x = 0$ Write in standard form.
$\qquad\qquad x(x - 8) = 0$
$x = 0$ or $x = 8$ Neither value is restricted.

The solution set is $\{0, 8\}$.

37. $$\frac{x}{x^2 - 9x + 14} - \frac{2}{x^2 - 6x + 8} = \frac{3}{x^2 - 11x + 28}$$

$$\frac{x}{(x - 7)(x - 2)} - \frac{2}{(x - 4)(x - 2)} = \frac{3}{(x - 7)(x - 4)}$$

$x \neq 2, 4, 7$

$\qquad x(x - 4) - 2(x - 7) = 3(x - 2)$ Multiply by the LCM.
$x^2 - 4x - 2x + 14 = 3x - 6$
$\qquad x^2 - 9x + 20 = 0$ Write in standard form. 4 is
$\qquad (x - 5)(x - 4) = 0$ a restricted value and must
$x = 5$ or $x = 4$ be rejected.

The solution set is $\{5\}$.

41. $$\frac{3}{x^2 + 4x - 21} - \frac{5x}{x^2 - 8x + 15} = \frac{8}{x^2 + 2x - 35}$$

$$\frac{3}{(x + 7)(x - 3)} - \frac{5x}{(x - 5)(x - 3)} = \frac{8}{(x + 7)(x - 5)}$$

$x \neq -7, 3, 5$

$3(x - 5) - 5x(x + 7) = 8(x - 3)$ Multiply by the LCM.
$3x - 15 - 5x^2 - 35x = 8x - 24$
$\qquad -5x^2 - 40x + 9 = 0$
$\qquad 5x^2 + 40x - 9 = 0$ Multiply by -1 so that the coefficient of x^2 is positive.

$a = 5, b = 40, x = -9$ Solve by using the quadratic formula.

$$x = \frac{-40 \pm \sqrt{1600 - 4(5)(-9)}}{10}$$

$$x = \frac{-40 \pm \sqrt{1780}}{10} = \frac{-40 \pm \sqrt{4}\sqrt{445}}{10}$$

$$x = \frac{-40 \pm 2\sqrt{445}}{10} = \frac{2(-20 \pm \sqrt{445})}{2(5)}$$

$$x = \frac{-20 \pm \sqrt{445}}{5}$$

Divide out the common factor, 2. Neither value is restricted.

The solution set is $\left\{\frac{-20 \pm \sqrt{445}}{5}\right\}$.

D

45. Jane and Sally each earned $600 last month. Sally is paid one dollar an hour more than Jane. If together they worked a total of 220 hours, how much is each paid per hour?

Select a variable:
Let x represent the pay per hour that Jane receives. So $x + 1$ represents Sally's pay per hour.

	Pay per hour	Earnings	Hours worked (Earnings/Pay per hour)
Jane	x	600	$\frac{600}{x}$
Sally	x + 1	600	$\frac{600}{x+1}$

Make a chart to organize the information.

Simpler word form:
$\begin{bmatrix} \text{Hours that} \\ \text{Jane works} \end{bmatrix} + \begin{bmatrix} \text{Hours that} \\ \text{Sally works} \end{bmatrix} = 220$

Translate to algebra:

$\frac{600}{x} + \frac{600}{x+1} = 220$

$600(x + 1) + 600(x) = 220(x)(x + 1)$ Multiply by the LCM.
$600x + 600 + 600x = 220x^2 + 220x$
$220x^2 - 980x - 600 = 0$ Simplify.
$11x^2 - 49x - 30 = 0$ Divide by 20.
$(11x + 6)(x - 5) = 0$ Factor.
$x = -\frac{6}{11}$ or $x = 5$ Reject $x = -\frac{6}{11}$ since x represents Jane's pay.

Check:

	Pay per hour	Earnings	Hours worked (Earnings/Pay per hour
Jane	5	600	$\frac{600}{5} = 120$
Sally	5+1=6	600	$\frac{600}{6} = 100$

Check in the original information chart. The total hours worked is 220.

Answer:
Jane earns $5 per hour and Sally earns $6 per hour.

E Maintain Your Skills

Solve:

49. $\dfrac{2x + 7}{14} - \dfrac{5x - 4}{3x + 1} = \dfrac{x + 6}{7}$

$(2x + 7)(3x + 1) - (5x - 4)(14) = (x + 6)(2)(3x + 1)$

                                                                                     Multiply by the LCM,
                                                                                     14(3x + 1).

$6x^2 + 23x + 7 - 70x + 56 = 6x^2 + 38x + 12$ Simplify.
$-47x + 63 = 38x + 12$
$-85x = -51$
$x = \dfrac{51}{85} = \dfrac{3(17)}{5(17)}$
$x = \dfrac{3}{5}$ Reduce.

Rationalize the denominator:

53. $\dfrac{2\sqrt{x} + \sqrt{y}}{\sqrt{x} - 2\sqrt{y}}$

$= \dfrac{(2\sqrt{x} + \sqrt{y})(\sqrt{x} + 2\sqrt{y})}{(\sqrt{x} - 2\sqrt{y})(\sqrt{x} + 2\sqrt{y})}$ Multiply both numerator and denominator by the conjugate of the denominator.

$= \dfrac{2x + 5\sqrt{xy} + 2y}{x - 4y}$ Simplify.

EXERCISES 5.9 EQUATIONS THAT ARE QUADRATIC IN FORM

A

Solve:

1. $x^4 - 3x^2 - 4 = 0$

   Let $u = x^2$ Assign new variable and write as a quadratic.
   $u^2 - 3u - 4 = 0$ Substitute.
   $(u - 4)(u + 1) = 0$ Factor and solve.
   $u = 4$ or $u = -1$ Solution of the quadratic equation but not the equation involving x.

   $u = 4$ or $u = -1$ Substitute $u = x^2$ and continue.
   $x^2 = 4$ $x^2 = -1$

   $x = \pm\sqrt{4}$ $x = \pm\sqrt{-1}$
   $x = \pm 2$ $x = \pm i$

The solution set is $\{\pm 2, \pm i\}$.

5. $x^4 + 2x^2 - 15 = 0$

    Let $u = x^2$ Assign new variable and write as a quadratic.

        $u^2 + 2u - 15 = 0$ Substitute.
       $(u + 5)(u - 3) = 0$ Factor.
        $u = -5$ or $u = 3$
        $x^2 = -5$ $x^2 = 3$ Substitute $u = x^2$ and continue.

        $x = \pm\sqrt{5}i$ $x = \pm\sqrt{3}$

    The solution set is $\{\pm\sqrt{3}, \pm\sqrt{5}i\}$.

9. $4x^4 - 25x^2 + 36 = 0$

    Let $u = x^2$
       $4u^2 - 25u + 36 = 0$ Substitute.
      $(4u - 9)(u - 4) = 0$ Factor.
        $u = \frac{9}{4}$ or $u = 4$

        $x^2 = \frac{9}{4}$ $x^2 = 4$ Substitute and continue.

        $x = \pm\sqrt{\frac{9}{4}}$ $x = \pm\sqrt{4}$

        $x = \pm\frac{3}{2}$ $x = \pm 2$

    The solution set is $\left\{\pm\frac{3}{2}, \pm 2\right\}$.

13. $(x - 5)^2 - 3(x - 5) - 10 = 0$

    Let $u = x - 5$
      $u^2 - 3u - 10 = 0$ Substitute.
    $(u + 2)(u - 5) = 0$ Factor.
        $u = -2$ or $u = 5$
    $x - 5 = -2$ or $x - 5 = 5$ Substitute $x - 5$ for u and
         $x = 3$ $x = 10$ continue, solving for x.

    The solution set is $\{3, 10\}$.

B

17. $(x^2 - 1)^2 - 4(x^2 - 1) + 3 = 0$

    Let $u = x^2 - 1$
       $u^2 - 4u + 3 = 0$ Substitute u for $x^2 - 1$.
    $(u - 3)(u - 1) = 0$ Factor.
          $u = 3$ or $u = 1$ Solve.
    $x^2 - 1 = 3$ $x^2 - 1 = 1$ Substitute $x^2 - 1$ for u,
        $x^2 = 4$ $x^2 = 2$ and continue solving for x.

        $x = \pm\sqrt{4}$ $x = \pm\sqrt{2}$
        $x = \pm 2$

    The solution set is $\{\pm\sqrt{2}, \pm 2\}$.

21. $8(x^2 - 2)^2 - 10(x^2 - 2) - 63 = 0$
Let $u = x^2 - 2$
$8u^2 - 10u - 63 = 0$
$(4u + 9)(2u - 7) = 0$
$4u + 9 = 0$ or $2u - 7 = 0$
$u = -\frac{9}{4}$ or $u = \frac{7}{2}$
$x^2 - 2 = -\frac{9}{4}$ or $x^2 - 2 = \frac{7}{2}$

$4x^2 - 8 = -9$ $\qquad$ $x^2 = \frac{7}{2} + 2 = \frac{7}{2} + \frac{4}{2}$
$4x^2 = -1$
$x^2 = -\frac{1}{4}$ $\qquad$ $x^2 = \frac{11}{2}$

$x = \pm\sqrt{-\frac{1}{4}}$ $\qquad$ $x = \pm\sqrt{\frac{11}{2}} \cdot \frac{\sqrt{2}}{\sqrt{2}}$

$x = \pm\frac{1}{2}\sqrt{-1}$ $\qquad$ $x = \pm\frac{\sqrt{22}}{2}$

$x = \pm\frac{1}{2}i$

The solution set is $\left\{\pm\frac{\sqrt{22}}{2}, \pm\frac{1}{2}i\right\}$.

25. $(x^2 - 10x)^2 = -11(x^2 - 10x) + 210$
Let $u = x^2 - 10x$
$u^2 = -11u + 210$ $\qquad$ Substitute u for $x^2 - 10x$.
$u^2 + 11u - 210 = 0$
$(u + 21)(u - 10) = 0$
$u = -21$ or $u = 10$ $\qquad$ Solve for u.
$x^2 - 10x = -21$ or $x^2 - 10x = 10$ $\qquad$ Substitute and solve for x.
$x^2 - 10x + 21 = 0$ $\qquad$ $x^2 - 10x - 10 = 0$ $\qquad$ Since $x^2 - 10x - 10$
$(x - 3)(x - 7) = 0$ $\qquad\qquad\qquad\qquad\qquad\qquad$ cannot be factored,
$x = 3$ or $x = 7$ $\qquad\qquad\qquad\qquad\qquad\qquad\quad$ solve for x by
$\qquad\qquad\qquad\qquad\qquad\qquad\qquad\qquad\qquad\qquad\quad$ using the quadratic
$\qquad\qquad\qquad\qquad\qquad\qquad\qquad\qquad\qquad\qquad\quad$ formula, where
$\qquad\qquad\qquad\qquad\qquad\qquad\qquad\qquad\qquad\qquad\quad$ $a = 1$, $b = -10$, and
$\qquad\qquad\qquad\qquad\qquad\qquad\qquad\qquad\qquad\qquad\quad$ $c = -10$.

$x = \frac{-b \pm \sqrt{b^2 - 4ac}}{2a}$

$= \frac{10 \pm \sqrt{100 - 4(1)(-10)}}{2}$

$= \frac{10 \pm \sqrt{140}}{2}$

$= \frac{10 \pm \sqrt{4 \cdot 35}}{2}$

$= \frac{10 \pm 2\sqrt{35}}{2} = \frac{2(5 \pm \sqrt{35})}{2}$

$= 5 \pm \sqrt{35}$

The solution set is $\{3, 7, 5 \pm\sqrt{35}\}$.

C

29. $x^{-2} - 5x^{-1} - 84 = 0$
Let $u = x^{-1}$
$u^2 - 5u - 84 = 0$ Substitute u for x^{-1}.
$(u - 12)(u + 7) = 0$
$u = 12$ or $u = -7$ Solve for u.
$x^{-1} = 12$ $x^{-1} = -7$ Substitute and solve for x.
$\frac{1}{x} = \frac{12}{1}$ $\frac{1}{x} = \frac{-7}{1}$
$1 = 12x$ $1 = -7x$
$\frac{1}{12} = x$ $-\frac{1}{7} = x$

The solution set is $\left\{-\frac{1}{7}, \frac{1}{12}\right\}$.

33. $2x - 13\sqrt{x} - 99 = 0$
Let $u = \sqrt{x}$
$2u^2 - 13u - 99 = 0$ If $u = \sqrt{x}$, then $u^2 = x$.
$(2u + 9)(u - 11) = 0$ Factor.
$2u + 9 = 0$ or $u - 11 = 0$
$u = -\frac{9}{2}$ or $u = 11$

$\sqrt{x} = -\frac{9}{2}$ or $\sqrt{x} = 11$ Substitute $\sqrt{x}$ for u and continue.
$x = 121$ Reject $\sqrt{x} = -\frac{9}{2}$, as the square root is positive and cannot equal a negative quantity.

The solution set is {121}.

37. $x^{-4} - 13x^{-2} + 36 = 0$
Let $u = x^{-2}$
$u^2 - 13u + 36 = 0$ If $u = x^{-2}$, then $u^2 = x^{-4}$.
$(u - 4)(u - 9) = 0$
$u - 4 = 0$ or $u - 9 = 0$
$u = 4$ or $u = 9$
$x^{-2} = 4$ $x^{-2} = 9$
$\frac{1}{x^2} = \frac{4}{1}$ $\frac{1}{x^2} = \frac{9}{1}$
$1 = 4x^2$ $1 = 9x^2$
$\frac{1}{4} = x^2$ $\frac{1}{9} = x^2$
$\pm\sqrt{\frac{1}{4}} = x$ $\pm\sqrt{\frac{1}{9}} = x$
$\pm\frac{1}{2} = x$ $\pm\frac{1}{3} = x$

The solution set is $\left\{\pm\frac{1}{3}, \pm\frac{1}{2}\right\}$.

D

41. The sum of the areas of two squares is 4160m^2. If the length of a side of the larger square is equal to the area of the smaller square, what is the length of each square?

Select variable:
Let x represent the length of a side of the smaller square.

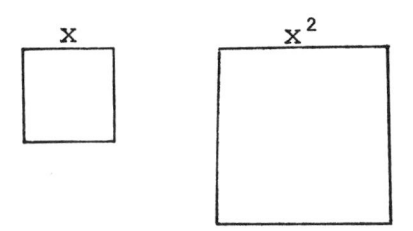

Organize the data by making a sketch. The area of the smaller square is x^2, so the length of a side of the larger square is x^2.

Simpler word form:
$$\begin{bmatrix} \text{Area of} \\ \text{smaller square} \end{bmatrix} + \begin{bmatrix} \text{Area of} \\ \text{larger square} \end{bmatrix} = 4160$$

Translate to algebra:
$x^2 + x^4 = 4160$

Solve:

$u^2 + u = 4160$

Let $u = x^2$. Substitute u for x^2, and u^2 for x^4.

$u^2 + u - 4160 = 0$
$(u + 65)(u - 64) = 0$
$u + 65 = 0$ or $u - 64 = 0$
$u = -65$ or $u = 64$
$x^2 = -65 x^2 = 64$
Reject. $ x = \pm 8$
$ x = 8$

Substitute x^2 for u.
Also, reject x = -8, since lengths are non-negative.

Check:

	Side	Area
Smaller Square	8	64
Larger Square	64	4096

64 + 4096 = 4160

The smaller square has sides of 8 m, and the sides of the larger square are each 64 m.

179

E Maintain Your Skills

Perform the indicated operations. Write answers in standard form:

45. $(7 + 3i) - (2 - 4i) + (5 - 6i)$
    $= 7 + 3i - 2 + 4i + 5 - 6i$
    $= 10 + i$

49. $(\sqrt{6} + 2i\sqrt{3})(\sqrt{2} + i\sqrt{3})$
    $= (\sqrt{6} \cdot \sqrt{2}) + (\sqrt{6} \cdot i\sqrt{3}) + (2i\sqrt{3} \cdot \sqrt{2}) + (2i\sqrt{3} \cdot i\sqrt{3})$
    $= \sqrt{12} + i\sqrt{18} + 2i\sqrt{6} + 2i^2\sqrt{9}$
    $= 2\sqrt{3} + 3i\sqrt{2} + 2i\sqrt{6} - 6$ $i^2 = -1$.

EXERCISES 5.10   ABSOLUTE VALUE INEQUALITIES

A

Solve and graph the solution:

1. $|x| < 4$
   $x < 4$ and $x > -4$
   or
   $-4 < x < 4$

   Write the absolute value inequality in its equivalent form.

   The graph of the solution is:

   ```
   ○————+——+——+——+——+——○
   -4         0         4
   ```

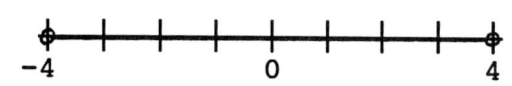

5. $|x| - 4 < -3$
   $|x| < 1$

   Simplify first by adding 4 to both sides.

   $x < 1$ and $x > -1$
   or
   $-1 < x < 1$

   Write the inequality in its equivalent form.

   The graph of the solution is:

   ```
   +——+——○————○——+——+
        -1  0  1
   ```

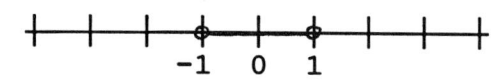

9. $|x| < -8$
   No solutions.

   Absolute value is always non-negative.

   The graph of the solution is:

   ```
   +——+——+——+——+——+——+
   -3 -2 -1  0  1  2  3
   ```

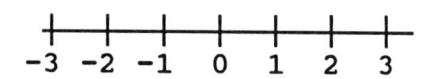

Solve, and write the solution in set builder notation:

13. $|5x| \geq 15$
    $5x \geq 15$ or $5x \leq -15$ Write the absolute value in
    $x \geq 3$ or $x \leq -3$ its equivalent form.

   In set builder notation, the solution is $\{x | x \leq -3 \text{ or } x \geq 3\}$.

B

Solve and write the solution in interval notation:

21. $|x + 2| - 3 < 4$
    $|x + 2| < 7$ Simplify by adding 3 to both sides.
    $x + 2 < 7$ and $x + 2 > -7$ Write and equivalent form for the inequality.
    $x < 5$ and $x > -9$ Solve each inequality.

   The solution is $(-9, 5)$.

25. $|x + 7| > -3$
    x can be any real number.   Since absolute value is always non-negative, the left side, being non-negative, is always greater than any negative quantity.

   The solution is $(-\infty, \infty)$.

29. $|2x + 3| \leq 5$
    $2x + 3 \leq 5$ and $2x + 3 \geq -5$ Write an equivalent form.
    $2x \leq 2$ $2x \geq -8$ Solve each inequality.
    $x \leq 1$ and $x \geq -4$

   The solution is $[-4, 1]$.

Solve and graph the solution:

33. $|3x + 5| < 2$
    $3x + 5 < 2$ and $3x + 5 > -2$ Write an equivalent form.
    $3x < -3$ $3x > -7$ Solve.
    $x < -1$ $x > -\frac{7}{3}$
    or
    $-\frac{7}{3} < x < -1$

   The graph of the solution is:

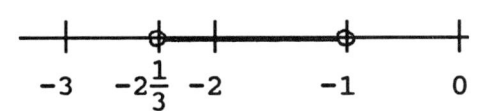

   ```
   ──┼────◦─┼──────◦─────┼───
    -3   -2⅓  -2       -1      0
   ```

181

C

37. $|2x + 4| > x - 5$
    $2x + 4 > x - 5$ or $2x + 4 < -(x - 5)$ Write an equivalent form for the inequality.
    $\quad x > -9$ or $\quad 2x + 4 < -x + 5$ Solve.
    $\qquad\qquad\qquad\qquad 3x < 1$
    $\qquad\qquad\qquad\qquad x < \frac{1}{3}$ Since either one or the other condition is always satisfied by any number, the solution set includes all real numbers.

    $x \in \mathbb{R}$

    The graph of the solution is:

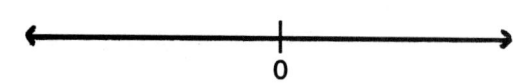

    <———————|———————>
              0

Solve and write the solution in set builder notation:

41. $|4x - 7| \geq 3x - 12$
    $4x - 7 \geq 3x - 12$ or $4x - 7 \leq -(3x - 12)$ Write the inequality in its equivalent form.
    $\quad x \geq -5 \qquad\qquad\quad 4x - 7 \leq -3x + 12$
    $\qquad\qquad\qquad\qquad\quad 7x \leq 19$ All real numbers make either the one or the other condition true, and so the solution set includes all real numbers.
    $\quad x \geq -5$ or $\qquad\quad x \leq \frac{19}{7}$

    The solution is $\{x | x \in \mathbb{R}\}$.

45. $|4 - x| \geq 4$
    $4 - x \geq 4$ or $4 - x \leq -4$ Write an equivalent form.
    $\;\; -x \geq 0 \qquad\quad\; -x \leq -8$
    $\;\;\;\; x \leq 0$ or $\qquad x \geq 8$

    The solution is $\{x | x \leq 0 \text{ or } x \geq 8\}$.

49. $|12 - 4x| - 8 \leq 12$
    $|12 - 4x| \leq 20$ Simplify first by adding 8 to both sides.
    $12 - 4x \leq 20$ and $12 - 4x \geq -20$ Write an equivalent form, and solve.
    $\;\; -4x \leq 8 \qquad\qquad -4x \geq -32$
    $\;\;\;\; x \geq -2$ and $\qquad\;\; x \leq 8$
    or
    $-2 \leq x \leq 8$

    The solution is $\{x | -2 \leq x \leq 8\}$.

53. $|2x + 1| > -x$

   $2x + 1 > -x$ or $2x + 1 < -(-x)$ Write an equivalent form.

   $3x > -1$ $2x + 1 < x$

   $x > -\frac{1}{3}$ or $x < -1$

   The solution is $\left\{x \mid x < -1 \text{ or } x > -\frac{1}{3}\right\}$.

D

In 57-64 write each as an absolute value inequality and solve:

57. The distance from a number and 4 is less than 12.

$\|x - 4\| < 12$	Use x to represent the number. The geometric meaning of absolute value is the distance between the number, x, and 4.
$x - 4 < 12$ and $x - 4 > -12$	Equivalent form.
$x < 16$ and $x > -8$	

   or
   $-8 < x < 16$

61. The distance from a number and 18 is at least 8.

$\|x - 18\|$	The geometric meaning of absolute value is the distance from x to 18.
$\|x - 18\| \geq 8$	"At least 8" is equivalent to "greater than or equal to 8."

   $x - 18 \geq 8$ or $x - 18 \leq -8$

   $x \geq 26$ or $x \leq 10$

65. The storage of food in a home freezer is best accomplished if the temperature in the freezer is maintained at 6°F plus or minus 8°. Find the range of permissible temperatures using absolute value inequalities.

$\|T - 6\| \leq 8$	The difference in the absolute value of the temperature in the freezer and 6°F must be less than or equal to 8°.
$T - 6 \leq 8$ and $T - 6 \geq -8$	Solve.
$T \leq 14$ and $T \geq -2$	

   or
   $-2 \leq t \leq 14$

   Thus, the inside temperatures should be between -2 and 14.

E   Maintain Your Skills

Factor completely:

69. $10xy - 15wy - 2x + 3w$
    $= 5y(2x - 3w) - (2x - 3w)$ Factor pairs.
    $= (5y - 1)(2x - 3w)$ Factor by grouping.

73. $g^3 - 9h^2g$
    $= g(g^2 - 9h^2)$ Factor out the common factor, g.
    $= g(g + 3h)(g - 3h)$ Difference of squares.

77. A cyclist climbs a three-mile-long hill at 40 miles per hour and returns at 48 miles per hour. What is his average rate? (Hint: average rate = total distance divided by total time.)

Average rate $= \dfrac{\text{Total distance}}{\text{Total time}}$ Formula.

	Distance ÷ Rate = Time		
Climb up	3	40	$\dfrac{3}{40}$
Return down	3	48	$\dfrac{3}{48}$

Make a chart to organize the data.

Total distance = 6 miles

Total time $= \dfrac{3}{40}$ hr $+ \dfrac{3}{48}$ hr

Average rate $= \dfrac{6}{\dfrac{3}{40} + \dfrac{3}{48}}$ Substitute using given formula.

$= \dfrac{6}{\dfrac{3}{40} + \dfrac{3}{48}} \cdot \dfrac{240}{240}$ Solve. Multiply by 1, using $\dfrac{240}{240}$, as the lowest common denomintor is 240.

$= \dfrac{6 \cdot 240}{\dfrac{3}{40} \cdot 240 + \dfrac{3}{48} \cdot 240}$

$= \dfrac{1440}{18 + 15}$

$= \dfrac{1440}{33}$

$= 43\dfrac{21}{33}$

$= 43\dfrac{7}{11}$

The average rate was $43\dfrac{7}{11}$ mph.

EXERCISES 5.11   QUADRATIC AND RATIONAL INEQUALITIES

A

Solve and graph. Write the solution in interval notation:

1. $(x + 2)(x - 5) > 0$

    The critical numbers are -2 and 5, shown below with vertical line.

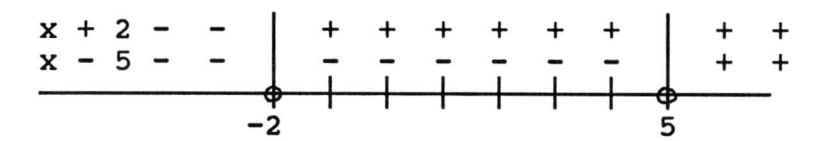

    ```
    x + 2  - -  |  + + + + + +  |  + +
    x - 5  - -  |  - - - - - -  |  + +
    ───────────⊖───────────────⊖────────
              -2                5
    ```

    Since $(x + 2)(x - 5) > 0$, the solution contains those numbers for which the factors are both negative and both positive.

    The solution is $(-\infty, -2) \cup (5, \infty)$.

5. $\dfrac{(5x + 3)}{(2x - 1)} \geq 0$ Standard form.

    $5x + 3 = 0 \qquad 2x - 1 = 0$
    $\quad x = -\dfrac{3}{5} \qquad\quad x = \dfrac{1}{2}$

    The critical values are $-\dfrac{3}{5}$ and $\dfrac{1}{2}$.

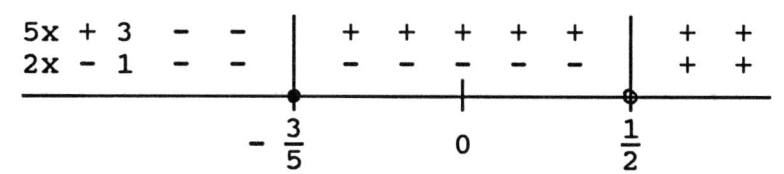

    ```
    5x + 3  - -  |  + + + + +  |  + +
    2x - 1  - -  |  - - - - -  |  + +
    ───────────●───────────────⊖────────
              -3/5         0         1/2
    ```

    Exclude the value $\dfrac{1}{2}$, since division by zero is undefined.

    Since the quotient is greater than or equal to 0, the solutions will be those numbers for which the factors are both negative, and those numbers for which the factors are both positive, including the critical value $-\dfrac{3}{5}$.

    The solution is $\left(-\infty, -\dfrac{3}{5}\right] \cup \left[\dfrac{1}{2}, \infty\right)$.

9. $-5x(x - 3)(x + 2) \geq 0$

   The critical numbers are 0, 3, and -2.

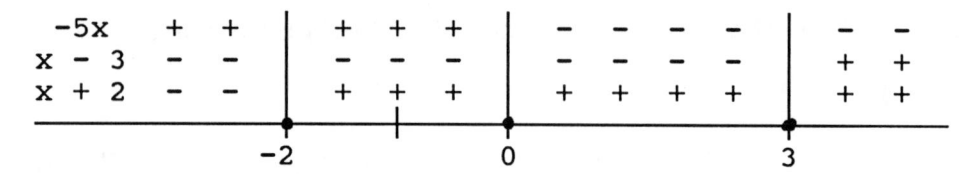

   ```
   -5x    + +  | + + +  | - - - -  | - -
   x - 3  - -  | - - -  | - - - -  | + +
   x + 2  - -  | + + +  | + + + +  | + +
   ─────●──────●─────────●──────
          -2         0         3
   ```

   The solution is $(-\infty, -2] \cup [0, 3]$.

13. $\dfrac{2x(2x + 3)}{(x - 5)} \geq 0$

   The critical numbers are 0, $-\dfrac{3}{2}$, and 5

   Set each factor equal to zero to determine critical numbers.

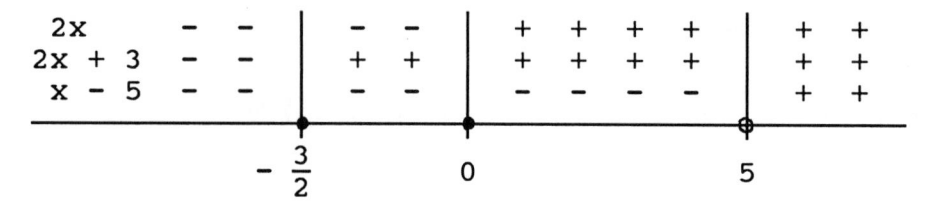

   ```
   2x      - -  | - -  | + + + +  | + +
   2x + 3  - -  | + +  | + + + +  | + +
   x - 5   - -  | - -  | - - - -  | + +
   ─────●──────●─────────○──────
         -3/2     0          5
   ```

   The solution is $\left[-\dfrac{3}{2}, 0\right] \cup (5, \infty)$.

B

17. $3x^2 + 5x + 4 > x^2 - 2x - 2$
    $2x^2 + 7x + 6 > 0$
    $(2x + 3)(x + 2) > 0$

    Standard form.
    Factor, and identify the critical numbers.

    The critical numbers are $-\dfrac{3}{2}$ and -2.

    ```
    2x + 3  - -  | - -  | + +
    x + 2   - -  | + +  | + +
    ─────○──────○──────
          -2      -3/2
    ```

    The solution is $(-\infty, -2) \cup \left(-\dfrac{3}{2}, \infty\right)$.

186

21. $\dfrac{2}{x} + 5 \geq 3$

$\dfrac{2}{x} + 2 \geq 0$ Simplify. Subtract 3 from both sides.

$\dfrac{2 + 2x}{x} \geq 0$ Write the left side with a common denominator, x.

$\dfrac{2(1 + x)}{x} \geq 0$

$\dfrac{1 + x}{x} \geq 0$ Divide both sides by 2.

The critical numbers are 0 and −1.   Set denominator and numerator equal to zero to find the critical numbers.

```
1 + x    - -  |  + +  |  + +
x        - -  |  - -  |  + +
──────────────●───────○──────────
             -1       0
```

The solution is $(-\infty, -1] \cup (0, \infty)$.

25. $x^2 + 5x + 10 < -5x - 15$

$x^2 + 10x + 25 < 0$ Standard form.

$(x + 5)^2 < 0$ Factor, to identtify critical numbers.

The critical number is −5.

```
x + 5    - -  |  + +
x + 5    - -  |  + +
──────────────○──────────
             -5
```

The solution contains numbers for which the factors have opposite signs (to produce a negative product). So the solution is the empty set, ∅.

29. $8x^2 + 20x + 15 \leq -8x^2 - 20x - 10$

$16x^2 + 40x + 25 \leq 0$ Standard form.

$(4x + 5)^2 \leq 0$

The critical number is $-\dfrac{5}{4}$.

```
4x + 25  - -  |  + +
4x + 25  - -  |  + +
──────────────●──────────
            -5/4
```

No numbers in either of the intervals have factors that are opposite in sign, so no negative products are produced. However, the critical number satisfies the equality.

The solution is $\left\{-\dfrac{5}{4}\right\}$.

C

Solve and graph. Write the solution in interval notation:

33. $$\frac{-4}{x + 5} > 4$$

$$\frac{-4}{x + 5} - 4 > 0 \qquad \text{Write the inequality in standard form.}$$

$$\frac{-4}{x + 5} - \frac{4(x + 5)}{x + 5} > 0$$

$$\frac{-4 - 4x - 20}{x + 5} > 0$$

$$\frac{-4x - 24}{x + 5} > 0 \qquad \text{Standard form.}$$

The critical numbers are -5 and -6.

```
-4x - 24   + +  | - - -  | - -
 x + 5     - -  | - - -  | + +
           ─────┼────────┼─────
                -6       -5
```

The solution set is (-6, -5).

The graph of the solution set is:

```
─────────────○━━━━━━━━━○─────────
             -6        -5
```

37. $$\frac{1}{x - 1} > \frac{3}{x - 2}$$

$$\frac{1}{x - 1} - \frac{3}{x - 2} > 0$$

$$\frac{1(x - 2) - 3(x - 1)}{(x - 1)(x - 2)} > 0$$

$$\frac{x - 2 - 3x + 3}{(x - 1)(x - 2)} > 0$$

$$\frac{-2x + 1}{(x - 1)(x - 2)} > 0 \qquad \text{Standard form.}$$

The critical numbers are $\frac{1}{2}$, 1, and 2.

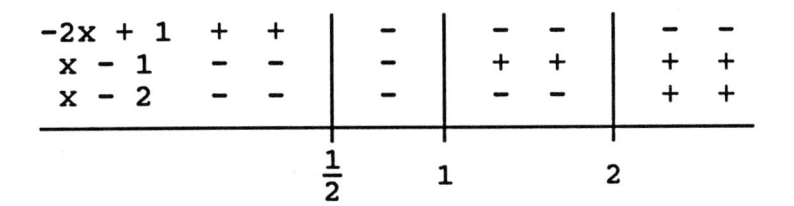

```
-2x + 1   + +  |  -  |  - -  |  - -
x - 1     - -  |  -  |  + +  |  + +
x - 2     - -  |  -  |  - -  |  + +
───────────────┼─────┼───────┼──────
              1/2    1       2
```

The solution set is $(-\infty, \frac{1}{2}) \cup (1, 2)$.

The graph of the solution set is:

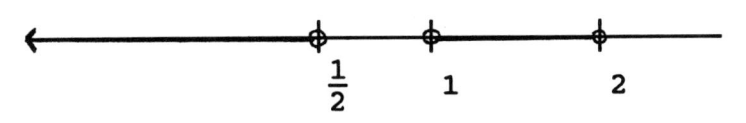

41.
$$\frac{1}{x-5} \geq \frac{1}{x+6}$$

$$\frac{1}{x-5} - \frac{1}{x+6} \geq 0$$

$$\frac{1(x+6) - 1(x-5)}{(x-5)(x+6)} \geq 0$$

$$\frac{x+6-x+5}{(x-5)(x+6)} \geq 0$$

$$\frac{11}{(x-5)(x+6)} \geq 0 \qquad \text{Standard form.}$$

The critical numbers are -6 and 5.

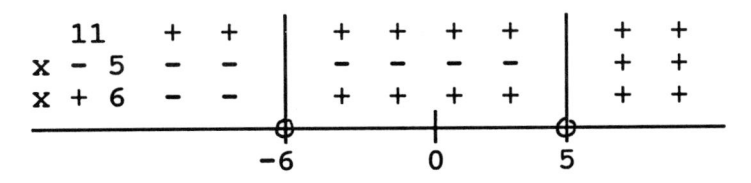

```
11      + +  | + + + +  | + +
x - 5   - -  | - - - -  | + +
x + 6   - -  | + + + +  | + +
─────────────┼──────────┼──────
            -6    0      5
```

The solution set is $(-\infty, -6) \cup (5, \infty)$.

The graph of the solution set is:

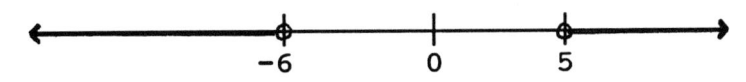

45.
$$\frac{x-5}{(x-1)(x+4)} \le 2$$

$$\frac{x-5}{(x-1)(x+4)} - 2 \le 0$$

$$\frac{x-5-2(x-1)(x+4)}{(x-1)(x+4)} \le 0$$

$$\frac{x-5-2(x^2+3x-4)}{(x-1)(x+4)} \le 0$$

$$\frac{x-5-2x^2-6x+8}{(x-1)(x+4)} \le 0$$

$$\frac{-2x^2-5x+3}{(x-1)(x+4)} \le 0 \quad \text{Standard form.}$$

$$\frac{(-2x+1)(x+3)}{(x-1)(x+4)} \le 0 \quad \text{Factor to identify critical numbers.}$$

The critical numbers are -4, -3, $\frac{1}{2}$, and 1.

```
-2x + 1   +  |  +  +  |  +  +  +  +  |  -     |  -
x + 3     -  |  -  -  |  +  +  +  +  |  +     |  +
x - 1     -  |  -  -  |  -  -  -  -  |  -     |  +
x + 4     -  |  +  +  |  +  +  +  +  |  +     |  +
             ⊖        •                •        ⊖
            -4       -3               1/2       1
```

The solution set is $(-\infty, -4) \cup \left[-3, \frac{1}{2}\right] \cup (1, \infty)$.

The graph of the solution is:

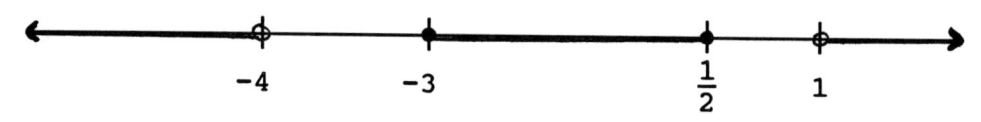

```
←————⊖————•————————•————⊖————→
    -4   -3       1/2   1
```

190

D

49. In an electronics factory, the cost (C) of producing x units of an item is given by $C = 8000 + 5x^2 - 350x$. If the cost is to be held at $12,000 or below, what is the range of units that can be produced?

$C = 8000 + 5x^2 - 350x$
or
$8000 + 5x^2 - 350x = C$ Symmetric property.
$8000 + 5x^2 - 350x \leq 12000$ The cost, C, is less than or equal to 12,000.
$5x^2 - 350x - 4000 \leq 0$ Standard form.
$5(x^2 - 70x - 800) \leq 0$ Factor.
$5(x - 80)(x + 10) \leq 0$

The critical numbers are -10 and 80.

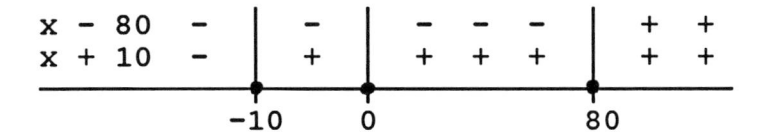

```
x - 80   -   |   -   |   - - -   |   + +
x + 10   -   |   +   |   + + +   |   + +
           ──●───────●───────────●──────
            -10      0          80
```

Although numbers between -10 and 0 result in factors whose products are below 12,000, they are nevertheless not included in the solution, as the number of units produced cannot be a negative number.
The solution set is therefore [0, 80].

The range of units that can be produced is 0 units to 80 units.

E   Maintain Your Skills

Simplify:

53. $25^{3/2} = (\sqrt{25})^3 = (5)^3 = 125$

57. $(81)^{1/4} = (\sqrt[4]{81})^1 = (3)^1 = 3$

Divide. Leave the answer in simplified form for radicals:

61. $\dfrac{\sqrt{3}}{1 - \sqrt{3}} = \dfrac{\sqrt{3}}{1 - \sqrt{3}} \cdot \dfrac{1 + \sqrt{3}}{1 + \sqrt{3}}$ Multiply by the conjugate of the denominator.

$\phantom{61.\ \dfrac{\sqrt{3}}{1 - \sqrt{3}}} = \dfrac{\sqrt{3} + 3}{1 - 3}$

$\phantom{61.\ \dfrac{\sqrt{3}}{1 - \sqrt{3}}} = \dfrac{\sqrt{3} + 3}{-2} \cdot \dfrac{-1}{-1}$ Multiply by -1 to clear the negative from the denominator.

$\phantom{61.\ \dfrac{\sqrt{3}}{1 - \sqrt{3}}} = \dfrac{-\sqrt{3} - 3}{2}$

CHAPTER 6

RELATIONS AND FUNCTIONS

EXERCISES 6.1 LINEAR EQUATIONS IN TWO VARIABLES

A

Find the ordered pairs that are solutions of the following equations that have the given values of x or y.

1. $2x + y = 5$

x	y
0	5

    $2(0) + y = 5$ Substitute 0 for x in the
    $0 + y = 5$ given equation. Solve for y.
    $y = 5$

5. $-2x + y = 4$

x	y
0	4

    $-2(0) + y = 4$ Substitute 0 for x in the
    $0 + y = 4$ given equation and solve
    $y = 4$ for y.

9. $-3x + 2y = 6$

x	y
-4	-3

    $-3x + 2(-3) = 6$ Substitute -3 for y in the
    $-3x - 6 = 6$ given equation. Solve
    $x = -4$ for x.

13. $5x - 2y = 10$

x	y
0	-5

    $5x - 2(-5) = 10$ Substitute -5 for y in the
    $5x + 10 = 10$ given equation, and solve
    $x = 0$ for x.

Draw the graph of each equation:

17. y = x - 2

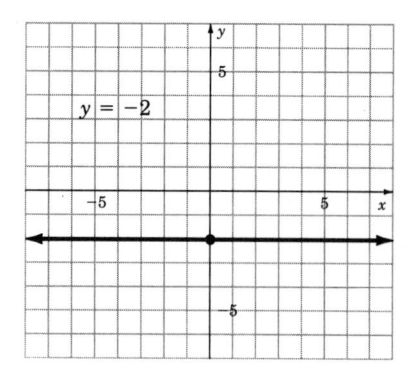

x	y
-2	-4
0	-2
2	0

Find a minimum of three points by assigning values to x and solving for y.

Plot each of the points on the coordinate system and draw the straight line that passes through the points.

21. y = -2

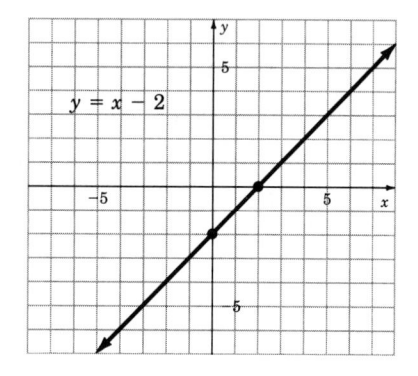

x	y
-2	-2
0	-2
1	-2

Find three points. Regardless of the value of x, the value of y is -2.

Plot the points and draw the line.

25. 2x + y = 3

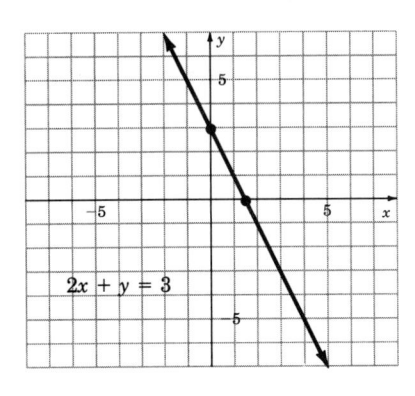

x	y
-2	7
0	3
3	-3

Assign values to x, and solve for y.

Plot the points and draw the line.

B

29. 7y - 6x = 42

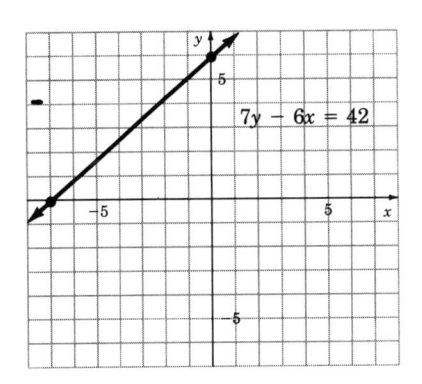

x	y
0	6
-7	0
-3	$-\dfrac{24}{7}$

Substitute assigned values for x, and solve for y.

Plot the points and draw the line.

33. x - 5y = -6

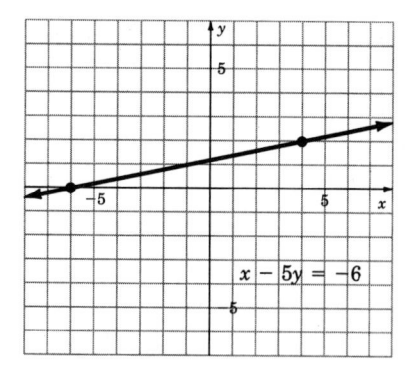

x	y
-6	0
0	6/5
4	2

Find three points.

Plot the points and draw the line.

37. 2x - 9y = 18

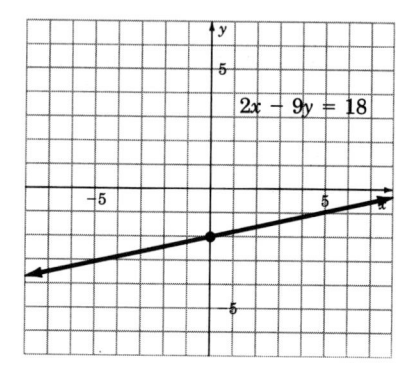

x	y
0	-2
3	-4/3
-3	-24/9

Find a minimum of three points.

Plot the points and draw the line.

196

41. $8x - y = 4$

x	y
0	-4
$\frac{1}{2}$	0
1	4

Find three points.

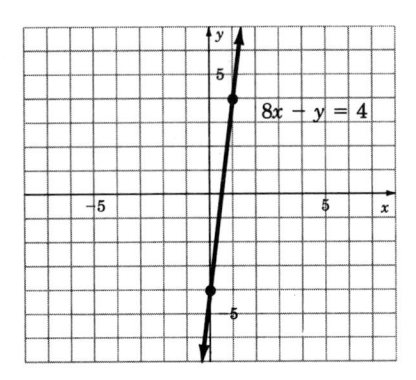

Plot each point and draw the line.

C

45. $y = \frac{3}{8}x + 6$

x	y
0	6
8	9
-8	3

Find three points.

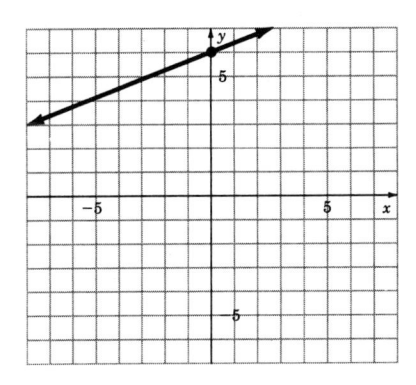

Plot the points and draw the line.

49. $\frac{1}{7}x + \frac{1}{4}y + 1 = 0$

x	y
-7	0
0	-4
3	$-\frac{40}{7}$

Find a minimum of three points.

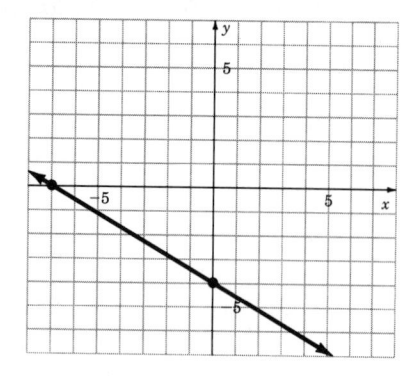

Draw the line connecting the points.

In Exercise 53, one unit on the x-axis equals 5, and one unit on the y-axis equals 10.

53. $25x - y = 125$

x	y
5	0
7	50
3	-50

Find a minimum of three points.

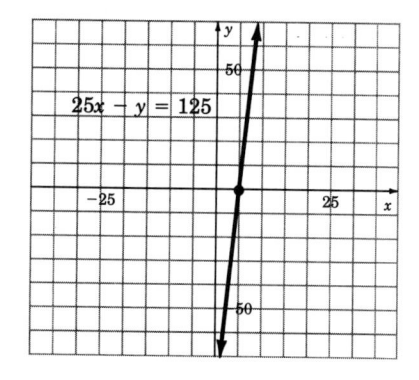

Plot each point and draw the line.

198

In exercise 57, let one unit on the x-axis equal 50 and one unit on the y-axis equal 25.

57. x - y = 200

x	y
75	-125
200	0
275	75

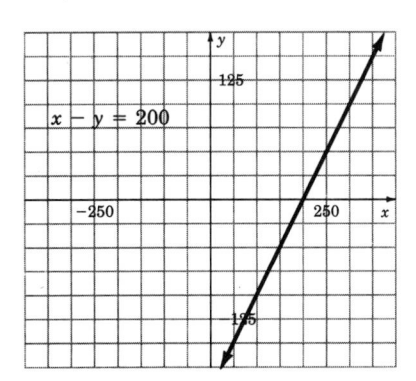

Find three points.

Plot the points and draw the graph.

61. A furniture refinishing outlet pays its employees $40 per day plus $10 per unit refinished. Write an equation to express an employee's daily wage (w) in terms of the units finished (x). Draw a graph for x ≥ 0.

w = 10x + 40   Let x represent the number of units finished. Then 10x represents the amount paid for x units. The base pay is 40, so add 40.

x	y
0	40
1	50
2	60

Find a minimum of three points.

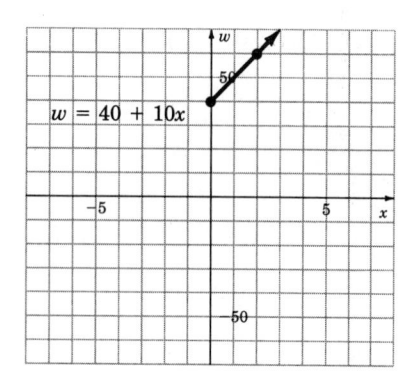

Since x ≥ 0, do not extend the line past the y-axis where x = 0.

E Maintain Your Skills

Combine and reduce, if possible:

65. $\dfrac{x + 1}{x - 6} + \dfrac{4}{x - 3} - \dfrac{2x - 4}{x^2 - 9x + 18}$

$= \dfrac{(x+1)(x-3)}{(x-6)(x-3)} + \dfrac{4(x-6)}{(x-6)(x-3)} - \dfrac{2x - 4}{(x-6)(x-3)}$

$= \dfrac{(x^2 - 2x - 3) + (4x - 24) - (2x - 4)}{(x - 6)(x - 3)}$

$= \dfrac{x^2 - 2x - 3 + 4x - 24 - 2x + 4}{(x - 6)(x - 3)}$

$= \dfrac{x^2 - 23}{(x - 6)(x - 3)}$

Since $x^2 - 9x + 18$ factors as $(x - 6)(x - 3)$, build each fraction to have a common denominator, $(x-6)(x-3)$.

Multiply.

Solve by factoring:

69. $y^2 - 7y - 18 = 0$
     $(y - 9)(y + 2) = 0$
     $y - 9 = 0$ or $y + 2 = 0$
     $y = 9$ or $y = -2$

Factor.
Zero-product property.

The solution set is $\{-2, 9\}$.

73. $4(3 + 4x) = 3x^2$
     $12 + 16x = 3x^2$
     $0 = 3x^2 - 16x - 12$
     $0 = (3x + 2)(x - 6)$
     $3x + 2 = 0$ or $x - 6 = 0$
     $3x = -2$
     $x = -\dfrac{2}{3}$ or $x = 6$

Factor.

The solution set is $\left\{-\dfrac{2}{3}, 6\right\}$.

EXERCISES 6.2   PROPERTIES OF GRAPHS OF LINEAR EQUATIONS I

A

Write the coordinates of the x- and y-intercepts:

1. $3x - 7y = 21$

    y-intercept   x-intercept
    $x = 0$ $y = 0$
    $3(0) - 7 = 21$ $3x - 7(0) = 21$ The intercepts occur at
    $y = -3$ $x = 7$ $x = 0$ and $y = 0$.

    $(0, -3)$ $(7, 0)$ Write the intercepts as
                                                ordered pairs.

    The x-intercept is $(7, 0)$, and the y-intercept is $(0, -3)$.

5. Is the graph of the line that passes through $(5, 8)$ and $(3, 10)$ parallel to the line that passes through $(9, 6)$ and $(4, 12)$?

    $m_1 = \dfrac{8 - 10}{5 - 3} = \dfrac{-2}{2} = -1$

    $m_2 = \dfrac{6 - 12}{9 - 4} = \dfrac{-6}{5} = -\dfrac{6}{5}$

    The slopes are not equal, so the lines are not parallel.

Find the slope of the line joining the following points:

9. $(4, 0), (0, -2)$

    $m = \dfrac{y_2 - y_1}{x_2 - x_1} = \dfrac{-2 - 0}{0 - 4}$ Formula for slope. Let
                                                                   $(x_2, y_2) = (0, -2)$ and
                                                                   $(x_1, y_1) = (4, 0)$.
    $= \dfrac{-2}{-4} = \dfrac{1}{2}$

13. Find the distance between the following pairs of points: $(1, 5)$, $(3, 5)$.

    $d = \sqrt{(x_2 - x_1)^2 + (y_2 - y_1)^2}$ Formula for distance.

    $d = \sqrt{(3 - 1)^2 + (5 - 5)^2}$ Substitute.

    $d = \sqrt{4 + 0}$ Simplify.

    $d = 2$

B

Write the coordinates of the x- and y-intercepts:

17. $2x - 3y = 11$

$y = 0$
$2x - 3(0) = 11$
$x = \frac{11}{2}$

$x = 0$
$2(0) - 3y = 11$
$y = -\frac{11}{3}$

Substitute $y = 0$, and solve for x. Then substitute $x = 0$, and solve for y.

The x-intercept is $\left(\frac{11}{2}, 0\right)$, and the y-intercept is $\left(0, -\frac{11}{3}\right)$.

Find the slope of the line joining the following points, and find the distance between the points:

21. $(-6, 3)$, $(8, -2)$

$m = \frac{3 - (-2)}{-6 - 8} = \frac{5}{-14} = -\frac{5}{14}$

$d = \sqrt{[8 - (-6)]^2 + (-2 - 3)^2}$
$= \sqrt{196 + 25}$
$= \sqrt{221}$

The slope is $-\frac{5}{14}$, and the distance between points is $\sqrt{221}$.

25. $(2, -6)$, $(5, -6)$

$m = \frac{-6 - (-6)}{2 - 5} = \frac{0}{-3} = 0$

$d = \sqrt{(2 - 5)^2 + [-6 - (-6)]^2}$
$= \sqrt{9 + 0}$
$= 3$

The slope is 0, and the distance is 3.

Are the line segments joining the following pairs of points perpendicular?

29. $(0, 3)$ and $(-2, 5)$
     $(7, 6)$ and $(9, 8)$

$m_1 = \frac{5 - 3}{-2 - 0} = \frac{2}{-2} = -1$

$m_2 = \frac{8 - 6}{9 - 7} = \frac{2}{2} = 1$

$m_1 \cdot m_2 = -1$

The product of the slopes is -1, therefore the line segments are perpendicular.

33. Determine if the line y - 2x = 1 is parallel or perpendicular to the line y = 2x + 5.

y - 2x = 1   y = 2x + 5

x	y
1	3
-1	-1

x	y
-1	3
-2	1

Find two points on the graph of each equation.

$m = \dfrac{-1 - 3}{-1 - 1} = \dfrac{-4}{-2} = 2$ $m = \dfrac{1 - 3}{-2 - (-1)} = \dfrac{-2}{-1} = 2$ Compute the slopes.

The slopes are the same.

For the equation y - 2x = 1, when x = 0, y = 1.
For the equation y = 2x + 5, when x = 0, y = 5.
Therefore, the y-intercepts are different.

Since the slopes are the same and the y-intercepts are different, the lines are parallel.

C

Write the coordinates of the x- and y-intercepts:

37. 2y - 7 = 0
     2y = 7
     $y = \dfrac{7}{2}$

Solve for y.
This is a horizontal line that passes through the point $\left(0, \dfrac{7}{2}\right)$. Therefore it has no x-intercept.

There is no x-intercept; the y-intercept is $\left(0, \dfrac{7}{2}\right)$.

Find the slope of the line joining the following pairs of points, and find the distance between the points:

41. $\left(-\frac{3}{7}, 4\right), \left(-\frac{1}{14}, \frac{3}{2}\right)$

$$m = \frac{\frac{3}{2} - 4}{-\frac{1}{14} - \left(-\frac{3}{7}\right)} = \frac{\frac{3}{2} - \frac{8}{2}}{-\frac{1}{14} + \frac{6}{14}} = \frac{-\frac{5}{2}}{\frac{5}{14}} = -7$$

$$d = \sqrt{\left[-\frac{1}{14} - \left(-\frac{3}{7}\right)\right]^2 + \left(\frac{3}{2} - 4\right)^2}$$

$$= \sqrt{\left(\frac{5}{14}\right)^2 + \left(-\frac{5}{2}\right)^2}$$

$$= \sqrt{\frac{25}{195} + \frac{25}{4}}$$

$$= \sqrt{\frac{25}{196} + \frac{1225}{196}}$$

$$= \sqrt{\frac{1250}{196}} = \sqrt{\frac{2 \cdot 625}{196}} = \frac{25}{14}\sqrt{2}$$

The slope is -7, and the distance between the points is $\frac{25}{14}\sqrt{2}$.

45. Determine if the line $2y + x = 3$ is parallel or perpendicular to the line $y = 2x + 1$.

$2y + x = 3$ $y = 2x + 1$

x	y
0	$\frac{3}{2}$
1	1

x	y
0	1
1	3

Find two points on the graph of each equation. Note that the y-intercepts are different.

$m = \frac{\frac{3}{2} - 1}{0 - 1} = \frac{\frac{1}{2}}{-1}$ $m = \frac{1 - 3}{0 - 1} = \frac{-2}{-1}$ Compute each slope.

$m = -\frac{1}{2}$ $m = 2$

$\left(-\frac{1}{2}\right)(2) = -1$ Find the product of the slopes.

Since the product of the slopes is -1, the lines are perpendicular.

Are the line segments joining the following pairs of points parallel, perpendicular, or neither?

49. $\left(\frac{2}{3}, -\frac{3}{4}\right)$ and $\left(\frac{1}{2}, -\frac{1}{4}\right)$

$\left(\frac{1}{6}, -\frac{1}{5}\right)$ and $\left(-\frac{31}{30}, -\frac{3}{5}\right)$

$m = \dfrac{-\frac{3}{4} - \left(-\frac{1}{4}\right)}{\frac{2}{3} - \frac{1}{2}} = \dfrac{-\frac{3}{4} + \frac{1}{4}}{\frac{4}{6} - \frac{3}{6}} = \dfrac{-\frac{1}{2}}{\frac{1}{6}}$ Compute the slope of each segment.

$= -3$

$m = \dfrac{-\frac{1}{5} - \left(-\frac{3}{5}\right)}{\frac{1}{6} - \left(-\frac{31}{30}\right)} = \dfrac{-\frac{1}{5} + \frac{3}{5}}{\frac{5}{30} + \frac{31}{30}} = \dfrac{\frac{2}{5}}{\frac{36}{30}}$

$m = \frac{1}{3}$

$(-3)\left(\frac{1}{3}\right) = -1$ Compute the product of the two slopes.

Since the product of the slopes is -1, the line segments are perpendicular.

D

53. The opposite sides of a parallelogram are parallel. Show that the points (-1, 5), (0, 9), (2, 2) and (1, -2) are vertices of a parallelogram.

$(x_1, y_1) = (-1, 5)$
$(x_2, y_2) = (1, -2)$
$m = \dfrac{-2 - 5}{1 - (-1)} = -\dfrac{7}{2}$ Parallel lines have the same slope. Compute the slopes.

$(x_1, y_1) = (0, 9)$
$(x_2, y_2) = (2, 2)$
$m = \dfrac{2 - 9}{2 - 0} = -\dfrac{7}{2}$ One pair of opposite sides are parallel, since the slopes are equal.

$(x_1, y_1) = (2, 2)$
$(x_2, y_2) = (1, -2)$
$m = \dfrac{-2 - 2}{1 - 2} = 4$ Check to see if the second pair of opposite sides are parallel by computing the slopes.

$(x_1, y_1) = (-1, 5)$
$(x_2, y_2) = (0, 9)$
$m = \dfrac{9 - 5}{0 - (-1)} = 4$ The slopes are equal.

Opposite sides are parallel, therefore the points are the vertices of a parallelogram.

57. Use the Pythagorean theorem to show that the points (-2, 3), (-2, -7), and (1, 3) are the vertices of a right triangle.

If the given points are the vertices of a right triangle, two pairs of points will be the endpoints of line segments that are perpendicular.

$(x_1, y_1) = (-2, 3)$
$(x_2, y_2) = (1, 3)$
$m = \dfrac{3 - 3}{1 - (-2)} = \dfrac{0}{3} = 0$

Compute the slopes.

Horizontal line.

$(x_1, y_1) = (-2, 3)$
$(x_2, y_2) = (-2, -7)$
$m = \dfrac{-7 - 3}{-2 - (-2)} = \dfrac{-10}{0}$

Both points have the same first coordinate, -2. Undefined. This segment has no slope, and is therefore vertical.

The given points are vertices of a right triangle.

Horizontal and vertical lines are perpendicular.

E Maintain Your Skills

Simplify. Assume that all variables represent positive numbers:

61. $\sqrt{120x^5y^8} = \sqrt{4x^4y^8}\sqrt{30x}$

$= 2x^2y^4\sqrt{30x}$

65. $\sqrt{48s^3t^3} - 3st\sqrt{75st} + t\sqrt{12s^3t}$

$= \sqrt{16s^2t^2}\sqrt{3st} - 3st\sqrt{25}\sqrt{3st} + t\sqrt{4s^2}\sqrt{3st}$

$= 4st\sqrt{3st} - 15st\sqrt{3st} + 2st\sqrt{3st}$

$= -9st\sqrt{3st}$

Solve by square roots:

69. $(a - 5)^2 = 40$

$\sqrt{(a - 5)^2} = \sqrt{40}$

$a - 5 = \pm 2\sqrt{10}$

$a = 5 \pm 2\sqrt{10}$

The solution set is $\{5 \pm 2\sqrt{10}\}$.

EXERCISES 6.3   PROPERTIES OF THE GRAPHS OF LINEAR EQUATIONS II

A

Write the equation of a line in standard form, given its slope and a point:

1. $(3, 6)$, $m = 4$
    $y - y_1 = m(x - x_1)$ Use the point-slope form of the equation.
    $y - 6 = 4(x - 3)$ Substitute.
    $y - 6 = 4x - 12$
    $6 = 4x - y$
    or $4x - y = 6$ Standard form.

5. $(-7, 8)$, $m = -\frac{1}{3}$
    $y - y_1 = m(x - x_1)$ Point-slope form.
    $y - 8 = -\frac{1}{3}[x - (-7)]$ Substitute.
    $y - 8 = -\frac{1}{3}(x - 7)$ Simplify.
    $3y - 24 = -x - 7$
    $x + 3y = 17$ Standard form.

Find the slope and the y-intercept of the graphs of the following equations. Draw the graph using the slope and the y-intercept.

9. $y = -x + 3$ Slope-intercept form.
    $m = -1$ and the y-intercept is $(0,3)$.

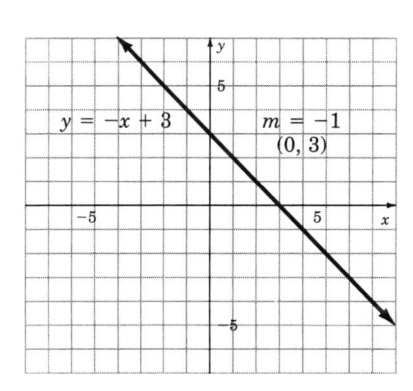

Plot the intercept. Move one unit right and one unit down. Draw the line passing through the points.

207

13. $y = -\frac{2}{3}x + 5$ Slope-intercept form.

$m = -\frac{2}{3}$ and the y-intercept is (0,5).

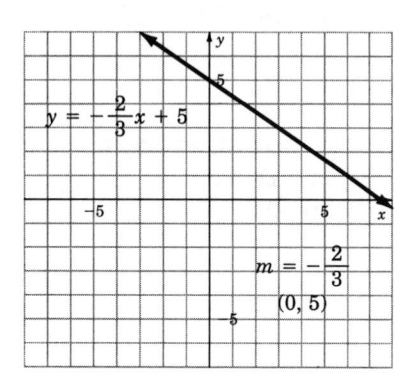

Plot the intercept. Move three units right and two units down.

B

Write the equation of the line in standard form, given two points on the line:

17. (6, -5), (8, -3)

$m = \frac{-3 - (-5)}{8 - 6} = \frac{2}{2} = 1$ Since the slope is not given, first calculate m.

$y - y_1 = m(x - x_1)$ Now use point-slope form with either given point.

$y - (-5) = 1(x - 6)$ Substitute the point (6, -5).
$y + 5 = x - 6$
$11 = x - y$
or $x - y = 11$ Standard form.

21. (-11, 12), (3, -5)

$m = \frac{-5 - 12}{3 - (-11)} = \frac{-17}{14} = -\frac{17}{14}$ First calculate m.

$y - (-5) = -\frac{17}{14}(x - 3)$ Next, substitute either point using the point-slope form. Here the point (3, -5) is substituted.

$y + 5 = -\frac{17}{14}(x - 3)$

$14y + 70 = -17x + 51$ Multiply both sides by 14.
$17x + 14y = -19$ Standard form.

Find the slope and the y-intercept of the graphs of the following equations. Draw the graphs using the slope and the y-intercepts:

25. $4x - y = 7$
     $y = 4x - 7$ Write in slope-intercept form.

   $m = 4$ and the y-intercept is $(0, -7)$.

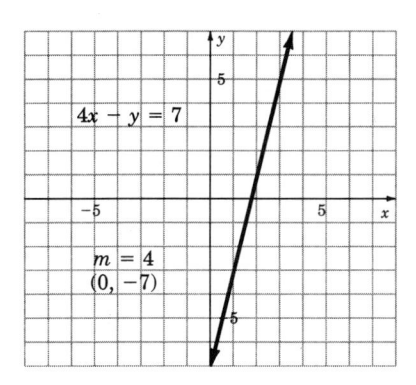

   Plot the y-intercept. Move one unit right and four units up.

29. $2x - 5y - 20 = 0$
     $-5y = -2x + 20$
     $y = \frac{2}{5}x - 4$

   $m = \frac{2}{5}$ and the y-intercept is $(0, -4)$.

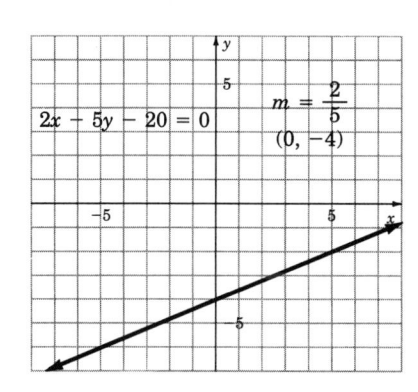

   Plot the y-intercept. Move five units right and two units up.

C

Write the equation of the line in standard form given two points on the line:

33. $\left[\frac{1}{2}, 4\right]$, $\left[-\frac{2}{3}, 4\right]$

$$m = \frac{4-4}{\frac{1}{2} - \left[-\frac{2}{3}\right]} = 0 \qquad \text{Calculate m.}$$

$$y - 4 = 0\left[x - \frac{1}{2}\right] \qquad \text{Substitute } \left[\frac{1}{2}, 4\right] \text{ using the point-slope formula.}$$

$$y - 4 = 0$$
$$y = 4$$

37. $\left[\frac{1}{3}, \frac{3}{4}\right]$, $\left[\frac{1}{2}, \frac{2}{3}\right]$

$$m = \frac{\frac{2}{3} - \frac{3}{4}}{\frac{1}{2} - \frac{1}{3}} = \frac{-\frac{1}{12}}{\frac{1}{6}}$$

$$= -\frac{1}{2}$$

$$y - \frac{3}{4} = -\frac{1}{2}\left[x - \frac{1}{3}\right] \qquad \begin{array}{l}\text{Using the point-slope form} \\ \text{substitute either point.} \\ \text{Here the point } \left[\frac{1}{3}, \frac{3}{4}\right] \text{ is} \\ \text{substituted.}\end{array}$$

$$12y - 9 = -6x + 2 \qquad \text{Simplify.}$$
$$6x + 12y = 11 \qquad \text{Multiply both sides by 12.}$$

Write the following equations in standard form and slope-intercept form.

41. $3(x - 1) - 5 = 2(y - 1) + 7$
$3x - 3 - 5 = 2y - 2 + 7$ Simplify.
$3x - 8 = 2y + 5$
$3x - 2y = 13$ Write in standard form.

$-2y = -3x + 13$ Now write in slope-intercept
$y = \frac{3}{2}x - \frac{13}{2}$ form by solving for y.

45. $4(x + 1) - 6 = 2(y + 4) + 5$
$4x + 4 - 6 = 2y + 8 + 5$ Simplify.
$4x - 2 = 2y + 13$
$4x - 2y = 15$ Write in standard form.

$-2y = -4x + 15$ Now write in slope-intercept
$y = 2x - \frac{15}{2}$ form.

D

49. The Uptown Corp. expects its profit to increase $1500 per year. If the profit in year three was $25,000, write the equation of the profit line. What will be the profit in year 12?

    Since the profit is $1500 per year, the slope is 1500.

Formula:
$y - y_1 = m(x - x_1)$

    Given the point $(x_1, y_1) = (3, 25{,}000)$ and the slope, $m = 1500$, use the point-slope form to find the equation.

Substitute:
$y - 25000 = 1500(x - 3)$

Solve:
$y - 25000 = 1500x - 4500$
$1500x - 4500 = y - 25000$
$1500x - y = -20500$ Standard form.

$1500(12) - y = -20500$
$-y = -20500 - 18000$
$y = 38500$

Evaluate the equation when $x = 12$.
Profit in year 12.

The standard form of the profit line is $1500x - y = -20500$, and the profit in year 12 will be $38,500.

53. Write the equation of a line that passes through the point $(-2, 5)$ and is parallel to the line whose equation is $2x - 3y = 5$.

$2x - 3y = 4$

$-3y = -2x + 4$

Parallel lines have the same slope. Write the given equation in the slope-intercept form to find the slope.

$y = \frac{2}{3}x - \frac{4}{3}$

So $m = \frac{2}{3}$

$y - y_1 = m(x - x_2)$

Use the given point and the slope to find the equation.

$y - 5 = \frac{2}{3}[(x - (-2))]$

$y - 5 = \frac{2}{3}(x + 2)$ Simplify.

$3y - 15 = 2x + 4$ Multiply both sides by 3.

$-19 = 2x - 3y$

or $2x - 3y = -19$

57. Write the equation of the line that passes through the point (2, -5) and is parallel to y + 3 = 0.

$$y + 3 = 0$$
$$y = -3$$ Solve for y.
$$y = 0x - 3$$ Slope-intercept form.
So m = 0.

$$y - y_1 = m(x - x_1)$$
$$y - (-5) = 0(x - 2)$$ Substitute the given point
$$y + 5 = 0$$ (2, -5) for (x_1, y_1) and
$$y = -5$$ replace m with 0.

E Maintain Your Skills

Multiply and simplify:

61. $(\sqrt{15} + \sqrt{24})(\sqrt{5} - \sqrt{6})$
   $= (\sqrt{15})(\sqrt{5}) + (\sqrt{15})(-\sqrt{6}) + (\sqrt{24})(\sqrt{5}) + (\sqrt{24})(-\sqrt{6})$
   $= \sqrt{75} - \sqrt{90} + \sqrt{120} - \sqrt{144}$
   $= 5\sqrt{3} - 3\sqrt{10} + 2\sqrt{30} - 12$ Simplify each radical.

Solve by completing the square:

65. $x^2 + 2x = 7$
   $x^2 + 2x + 1 = 7 + 1$ Add $\left[\frac{1}{2} \cdot 2\right]^2$ to both sides.
   $(x + 1)^2 = 8$ Factor on the left. Simplify on the right.

   $x + 1 = \pm\sqrt{8}$
   $x + 1 = \pm 2\sqrt{2}$
   $x = -1 \pm 2\sqrt{2}$

The solution set is $\{-1 \pm 2\sqrt{2}\}$.

EXERCISES 6.4   LINEAR INEQUALITIES

A

Draw the graph of each inequality:

1. y > x + 1
    y = x + 1

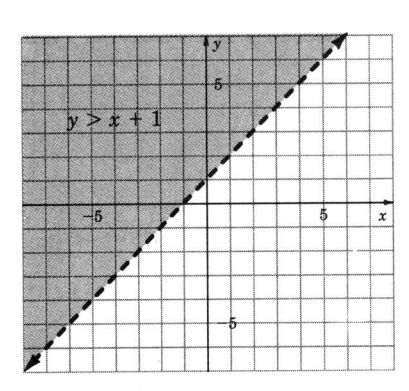

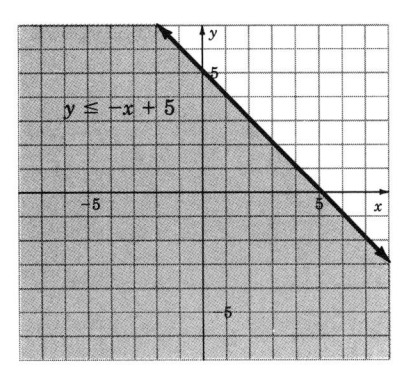

    Draw the graph of the corresponding equality.

x	y
0	1
-1	0
1	2

    The graph of the line is broken since equality is not included. Test the point (0, 0). 0 > 0 + 1 False. The graph is the half plane not containing the origin.

5. y ≤ -x + 5
    y = -x + 5

    Draw the graph of the equality.

x	y
0	5
5	0
2	3

    The graph of the line is solid since the relation includes equality. Test the point (0, 0).
    0 ≤ 0 + 5 True.
    The graph is the half plane containing the origin.

213

9. $x > -3$
    $x = -3$

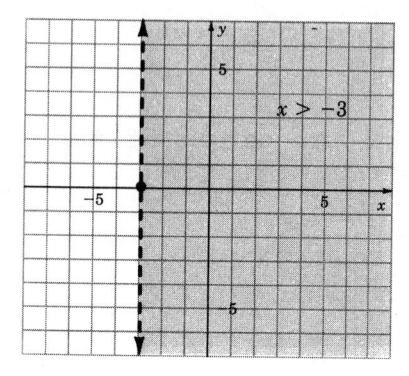

Vertical line with x-intercept at (-3, 0). Draw the graph of the equality.

The graph of the line is broken since equality is not included. Test the point (0, 0).
$0 > -3$ True.
The graph is the half-plane containing the origin.

B

13. $x - y < 4$
    $x - y = 4$

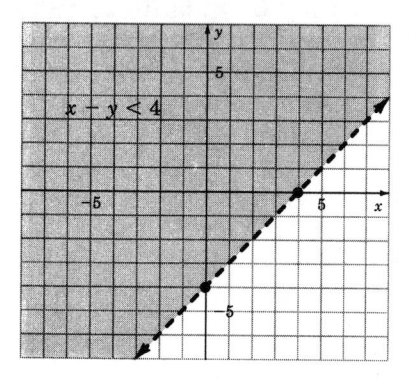

Draw the graph of the corresponding equality.

x	y
0	-4
4	0
2	-2

Test the point (0, 0).
$0 - 0 < 4$ True.
The graph is the half-plane containing the origin.

17. $3x - 5y < 10$
    $3x - 5y = 10$

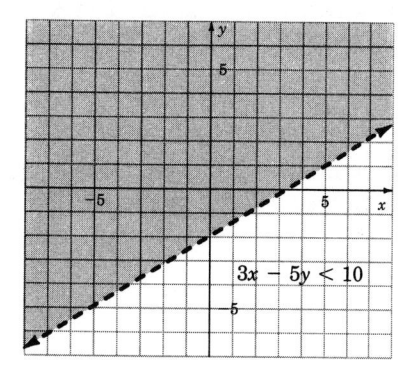

Draw the graph of the corresponding equality.

x	y
0	-2
$\frac{10}{3}$	0
-5	-5

The graph of the line is broken since equality is not included. Test the point (0, 0).
$3(0) - 5(0) < 10$
         $0 < 10$ True.
The graph is the half-plane containing the origin.

C

21. $2x - 7y \leq 14$
    $2x - 7y = 14$

Draw the graph of the equality.

x	y
0	-2
7	0
3	$-\frac{8}{7}$

The graph of the line is solid since equality is included in the relaion. Test the point (0, 0).
$0 - 0 \leq 14$ True.
The graph of the inequality is the half-plane containing the origin.

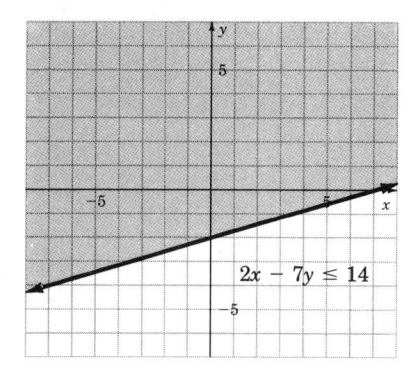

25. $\frac{1}{2}y + \frac{1}{2} - x \leq 0$
    $\frac{1}{2}y + \frac{1}{2} - x = 0$

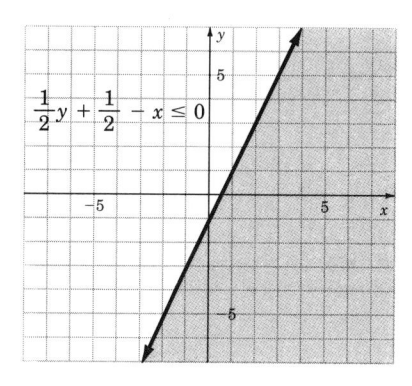

Draw the graph of the equality.

x	y
0	-1
$\frac{1}{2}$	0
2	3

The graph of the line is solid. Test the point (0, 0). $0 + \frac{1}{2} - 0 \leq 0$ False.
The graph of the inequality is the half-plane not containing the origin.

D

29. A dump truck with a divided bed can haul two different kinds of barkdust. Ground cover barkdust is heavier than decorator barkdust. The truck can carry up to 40 cubic yards at most. Draw the graph that shows the possible loads the truck can haul.

Simpler word form:

$$\begin{bmatrix} \text{Cubic yards} \\ \text{of ground} \\ \text{cover} \end{bmatrix} + \begin{bmatrix} \text{Cubic yards} \\ \text{of decorator} \\ \text{barkdust} \end{bmatrix} \leq 40$$

Translate to algebra:
x + y ≤ 40

The total number of cubic yards of barkdust cannot exceed 40. Draw the graph of the inequality.
Note that the graph is not valid for negative values of x or y. Test the point (10, 10). 10 + 10 ≤ 40 True.

Graph:

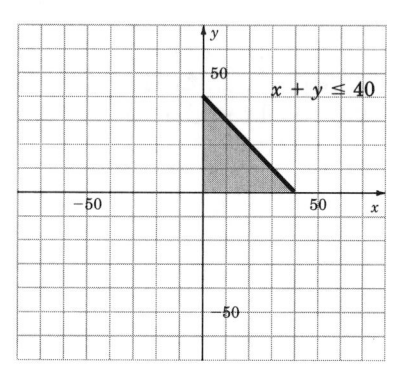

Any point on the line or in the shaded region has coordinates that represent a possible load for the truck.

33. John's math test consists of 50 true-false questions and 50 multiple-choice problems. Each true-false question is worth 5 points. Draw the graph that shows the combinations of multiple-choice and true-false questions he needs to answer correctly to receive a score of at least 75.

Simpler word form:

$$3\begin{bmatrix} \text{Number of} \\ \text{true-false questions} \\ \text{answered correctly} \end{bmatrix} + 5\begin{bmatrix} \text{Number of} \\ \text{multiple-choice questions} \\ \text{answered correctly} \end{bmatrix} \geq 75$$

Select variables:
Let x represent the number of true-false questions John answers correctly and y represent the number of multiple-choice questions he answers correctly.

Translate to algebra:
3x + 5 ≥ 75

<u>Solve:</u>
3x + 5y = 75

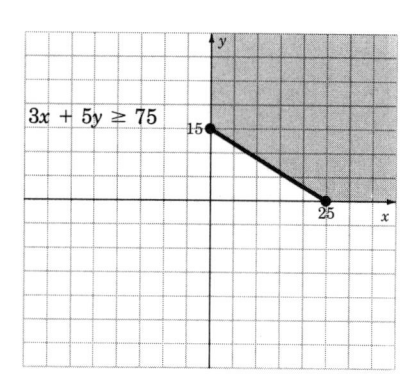

Draw the graph of the equality.

x	y
0	15
25	0
10	9

The graph of the line is solid since equality is included. Test the point (0, 0). 0 + 0 ≥ 75 False. The graph of the inequality is the half-plane not containing the origin. Notice that the number of questions answered correctly is greater than or equal to zero, so the negative values for x and y are not included in the graph.

Any point on the line or in the shaded region has coordinates that represent combinations of multiple-choice and true-false questions John needs to answer correctly to receive a score of at least 75.

E Maintain Your Skills

Divide (rationalize the denominator) and simplify:

37. $\dfrac{\sqrt{3}}{\sqrt{6}} - \dfrac{\sqrt{3}}{\sqrt{15}}$

$= \dfrac{\sqrt{3}}{\sqrt{6}} \cdot \dfrac{\sqrt{6}}{\sqrt{6}} - \dfrac{\sqrt{3}}{\sqrt{15}} \cdot \dfrac{\sqrt{15}}{\sqrt{15}}$

$= \dfrac{\sqrt{18}}{6} - \dfrac{\sqrt{45}}{15}$ Multiply.

$= \dfrac{3\sqrt{2}}{6} - \dfrac{3\sqrt{5}}{15}$ Simplify.

$= \dfrac{\sqrt{2}}{2} - \dfrac{\sqrt{5}}{5}$ Reduce.

$\dfrac{\sqrt{2}}{2} \cdot \dfrac{\sqrt{5}}{5} - \dfrac{\sqrt{5}}{5} \cdot \dfrac{2}{2}$ Build each fraction to have a common denominator, 10.

$\dfrac{5\sqrt{2} - 2\sqrt{5}}{10}$

Solve by the quadratic formula:

41. $4x^2 + 29x - 63 = 0$
    $a = 4$, $b = 29$, and $c = -63$

$x = \dfrac{-b \pm \sqrt{b^2 - 4ac}}{2a}$ Quadratic formula.

$x = \dfrac{-29 \pm \sqrt{29^2 - 4(4)(-63)}}{2(4)}$ Substitute.

$= \dfrac{-29 \pm \sqrt{841 + 1008}}{8}$

$= \dfrac{-29 \pm \sqrt{1849}}{8}$

$= \dfrac{-29 \pm 43}{8}$

$x = \dfrac{-29 + 43}{8}$ or $x = \dfrac{-29 - 43}{8}$

$x = \dfrac{14}{8}$ $x = \dfrac{-72}{8}$

$x = \dfrac{7}{4}$ or $x = -9$

The solution set is $\left\{-9, \dfrac{7}{4}\right\}$.

EXERCISES 6.5   RELATIONS

A

State the domain and range of each relation:

1. $\{(1, 1), (2, 4), (3, 9), (4, 16)\}$

   Domain: $\{1, 2, 3, 4\}$ The domain is the set of first numbers of the ordered pairs.

   Range: $\{1, 4, 9, 16\}$ The range is the set of second numbers of the ordered pairs.

5. $\{(-3, 8), (4, 8), (5, 8), (0, 8), (1, 8)\}$

   Domain: $\{-3, 0, 1, 4, 5)$ List in order the first numbers of the ordered pairs.

   Range: $\{8\}$ All the ordered pairs have 8 as the second number.

Write the following relation as a set of ordered pairs and state the range:

9. $\{(x, y) | y = |5x - 6| + 4, x = -3, -1, 1, 3\}$

x	y = |5x - 6| + 4		
-3	y = |-15 - 6| + 4 = 25	(-3, 25)	
-1	y = |-5 - 6| + 4 = 15	(-1, 15)	
1	y = | 5 - 6| + 4 = 5	(1, 5)	
3	y = |15 - 6| + 4 = 13	(3, 13)	

Substitute the given values of x in the equation to find the corresponding values of y. Form the ordered pairs.

$\{(-3, 25), (-1, 15), (1, 5), (3, 13)\}$

Range: $\{5, 13, 15, 25\}$

The relation as a set of ordered pairs. The range is the set of second components in the ordered pairs.

Draw the graph of the following and state the domain and range:

13. $y = 2\sqrt{x}$

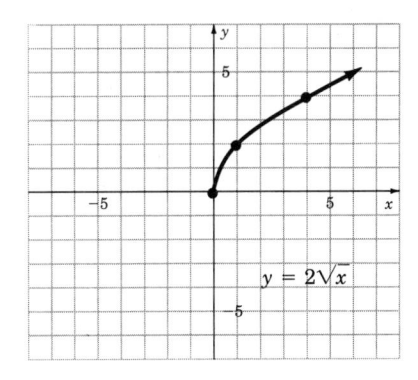

Draw the graph of the relation.

x	y
0	0
1	2
4	4

Domain: $\{x | x \geq 0\}$
Range: $\{y | y \geq 0\}$

B

Write each of the following relations as a set of ordered pairs and state the range:

17. $\{(x, y) \mid y = \sqrt{100 - x^2}, x = 0, \pm 4, \pm 5, \pm 8, \pm 10\}$

x	$y = \sqrt{100 - x^2}$		(x, y)
0	$y = \sqrt{100 - 0}$	= 10	(0, 10)
±4	$y = \sqrt{100 - 16}$	= $2\sqrt{21}$	(±4, $2\sqrt{21}$)
±5	$y = \sqrt{100 - 25}$	= $5\sqrt{3}$	(±5, $5\sqrt{3}$)
±8	$y = \sqrt{100 - 64}$	= 6	(±8, 6)
±10	$y = \sqrt{100 - 100}$	= 0	(±10, 0)

Substitute the given values of x in the equation. Form the ordered pairs.

$\{(0, 10), (\pm 4, 2\sqrt{21}), (\pm 5, 5\sqrt{3}), (\pm 8, 6), (\pm 10, 0)\}$

Range: 10, $2\sqrt{21}$, $5\sqrt{3}$, 6, 0

The range is the set of second components in the ordered pairs.

21. $\{(x, y) \mid y^2 = 3x^2 + 1, x = -2, -1, 0, 1, 2\}$

x	$y^2 = 3x^2 + 1$	(x, y)
-2	$y^2 = 3(4) + 1 = 13$ $y = \pm\sqrt{13}$	(-2, $\pm\sqrt{13}$)
-1	$y^2 = 3(1) + 1 = 4$ $y = \pm 2$	(-1, ±2)
0	$y^2 = 3(0) + 1 = 1$ $y = \pm 1$	(0, ±1)
1	$y^2 = 3(1) + 1 = 4$ $y = \pm 2$	(1, ±2)
2	$y^2 = 3(4) + 1 = 13$ $y = \pm\sqrt{13}$	(2, $\pm\sqrt{13}$)

Substitute the given values of x in the equation to find the corresponding values of y. Form the ordered pairs.

$\{(\pm 2, \pm\sqrt{13}), (\pm 1, \pm 2), (0, \pm 1)\}$
Range: $\{\pm\sqrt{13}, \pm 2, \pm 1\}$

State the domain and range of the following relations $x \in \mathbb{R}, y \in \mathbb{R}$.

25. $\{(x, y) | y = |x| - 7\}$

    Domain: $\{x | x \in \mathbb{R}\}$ There is no restriction for x since absolute value of any x is defined.

    Range: $\{y | y \geq -7, y \in \mathbb{R}\}$ The least value of $|x|$ is 0 since absolute value of negative numbers is positive. So $y \geq |0| - 7$ and $y \geq -7$

29. $\{(x, y) | y = \sqrt{x} + 6\}$

    Domain: $\{| x \geq 0, x \in \mathbb{R}\}$ The $\sqrt{x}$ is undefined for negative values of x.

    Range: $\{y | y \geq 6, y \in \mathbb{R}\}$ The least value of $\sqrt{x}$ is 0, so $y \geq \sqrt{0} + 6$ and $y \geq 6$

33. $\{(x, y) | y = -2x^2 + 1\}$

    Domain: $\{x | x \in \mathbb{R}\}$ There is no restriction for x since the square of x is defined for all x.

    Range: $\{y | y \leq 1\}$ The least value for x^2 is 0, since x^2 is always non-negative. So $y = -2(0) + 1$ and $y = 1$. All values of y for $x > 0$ are less than 1, since the form $-2x^2$ is always negative for those x-values.

C

Draw the graph of each of the following and state the domain and range:

37. $y = \sqrt{x - 4}$

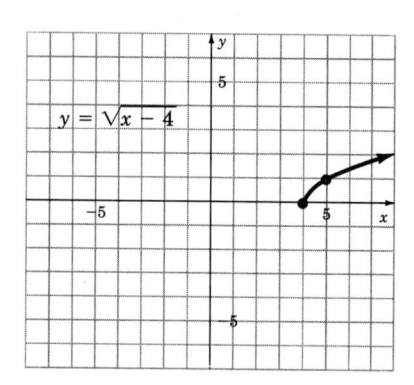

Make a table of values:

x	y
4	0
5	1
13	3

Domain: $\{x \mid x \geq 4, x \in \mathbb{R}\}$

Range: $\{y \mid y \geq 0, y \in \mathbb{R}\}$

From the graph.

41. $y = 2x^2 + 1$

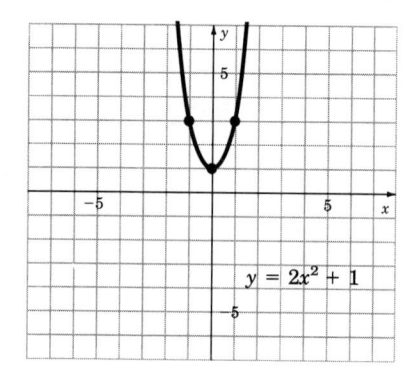

Make a table of values:

x	y
0	1
±1	3
±2	9

Plot the points and draw a smooth curve. The graph is a parabola.

Domain: $\{x \mid x \in \mathbb{R}\}$

Range: $\{y \mid y \geq 1\}$

From the graph.

State the domain and range of the following relations, $x \in \mathbb{R}$, $y \in \mathbb{R}$.

222

45. $\left\{(x, y) \mid y = \dfrac{1}{x - 1}\right\}$

    Domain: $\{x \mid x \neq 1, x \in \mathbb{R}\}$

    Range: $\{y \mid y \neq 0, y \in \mathbb{R}\}$

Restrict $x \neq 1$ since division by zero is undefined.
A fraction can equal zero only if the numerator equals zero. Here the numerator equals one, so the fraction can never equal zero.

49. $\left\{(x, y) \mid y = \dfrac{1}{2}x^2 - 2\right\}$

    Domain: $\{x \mid x \in \mathbb{R}\}$

    Range: $\{y \mid y \geq -2\}$

There is no restriction for x, since x^2 is always defined.
The quantity x^2 is always either 0, or positive. If $x = 0$, then $y = -2$. For all other values of x, a positive quantity would be added to -2 so that $y \geq -2$ for any x.

Draw the graph of each of the following and state the domain and range:

53. $y = \dfrac{1}{3}x^2 + 3$

Make a table of values:

x	y
0	3
±3	6
±1	$\dfrac{10}{3}$

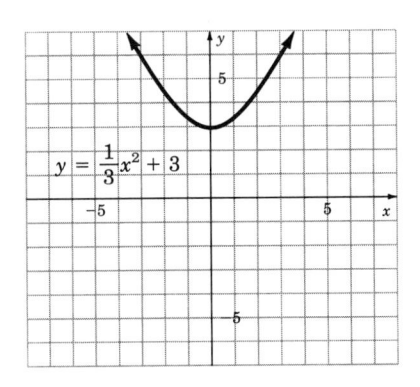

Plot the points and connect them with a smooth curve.

    Domain: $\{x \mid x \in \mathbb{R}\}$

    Range: $\{y \mid y \geq 3, y \in \mathbb{R}\}$

From the graph.

D

57. During the growing season, the height (h) (in feet) of an elephant-ear bamboo is related to the number of good growing days (d) by the formula $h = \sqrt{\frac{1}{2}d^2 + 4}$. Find the height of the bamboo after 4, 6, 10, and 15 growing days. Express the answers as ordered pairs, rounding the height to the nearest tenth of a foot.

d	h
4	3.5
6	4.7
10	7.3
15	10.8

Make a table of values to generate the ordered pairs (d, h).

The ordered pairs are (4, 3.5), (6, 4.7), (10, 7.3) and (15, 10.8).

61. The arch of a bridge span is a semicircle. The equation representing the arch is $x^2 + y^2 = 16$, $y \geq 0$. Draw the graph representing the arch.

x	y
0	4
±4	0
±1	$\sqrt{15}$

Make a table of values. Remember to restrict the values of y so that $y \geq 0$.

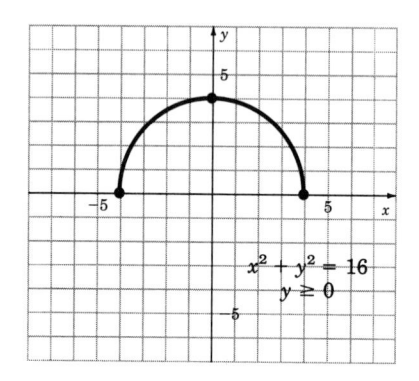

Plot the points and connect them with a smooth curve. The graph is a semi-circle.

224

E Maintain Your Skills

Solve:

65. $\dfrac{2}{3x} - \dfrac{3}{4x} - \dfrac{4}{5x} - \dfrac{5}{6x} = \dfrac{1}{30}$

$\qquad 2(20) - 3(15) - 4(12) - 5(10) = 2x$ Multiply each term by the
$\qquad \qquad \qquad 40 - 45 - 48 - 50 = 2x$ LCM of the denominators,
$\qquad \qquad \qquad \qquad \qquad \qquad -103 = 2x$ 60x.
$\qquad \qquad \qquad \qquad \qquad \quad -\dfrac{103}{2} = x$

The solution set is $\left\{-\dfrac{103}{2}\right\}$.

69. $\qquad \qquad \sqrt{x} + 4 = \sqrt{7}$

$\quad x + 8\sqrt{x} + 16 = 7$ Square both sides.
$\qquad \qquad 8\sqrt{x} = -x - 9$ Isolate the term with the radical.
$\qquad \qquad 8\sqrt{x} = -(x + 9)$ Factor. Since x is restricted to values greater than or equal to zero

(because of $\sqrt{x}$ on the left), and since for those values of x, the right side is always negative, this equation represents a contradiction.

The solution set is the empty set, ∅.

EXERCISES 6.6   FUNCTIONS

A

Identify whether the following relations are functions:

1. R = {(-2, 1), (-1, 2), (0, 3), (1, 4), (2, 5)}

   It is a function.   No two ordered pairs have the same x value.

5. $\left\{(x, y) \mid y - \dfrac{1}{2}x\right\}$

   It is a function.   For each value of x there is exactly one value of y.

Find the indicated values of f(x):

9. $f(x) = 9x - 7$; $f(x)$, $f(-3)$

   $f(2) = 9(2) - 7 = 11$ Replace x by 2.
   $f(-3) = 9(-3) - 7 = -34$ Replace x by -3.

   So $f(2) = 11$ and $f(-3) = -34$.

13. $f(x) = x^2 + 2$; $f(0)$, $f(-1)$

    $f(0) = 0^2 + 2 = 2$ Replace x by 0.
    $f(-1) = (-1)^2 + 2 = 3$ Replace x by -1.

    So $f(0) = 2$ and $f(-1) = 3$

B

Identify whether the following relations are functions:

17. $y = |x + 3| - 2$

x	y
2	3
3	4
-8	3
-9	4

(2, 3)
(3, 4)
(-8, 3)
(-9, 4)

List some ordered pairs.

Form the ordered pairs.

It is a function.

For each value of x there is exactly one value of y. In the ordered pairs, all x values are different.

21. $y = \sqrt{x - 9}$

x	y
9	0
18	3
45	6

(9, 0)
(18, 3)
(45, 6)

Form ordered pairs.

It is a function.

For each value of x there is exactly one value of y. In the ordered pairs, all x values are different.

Find the indicated values of f(x):

25. $f(x) = \frac{4}{x + 5}$; $f(-3)$, $f(6)$, $f(0)$

$f(-3) = \frac{4}{-3 + 5} = \frac{4}{2} = 2$

$f(6) = \frac{4}{6 + 5} = \frac{4}{11}$

$f(0) = \frac{4}{0 + 5} = \frac{4}{5}$

So $f(-3) = 2$, $f(6) = \frac{4}{11}$, and $f(0) = \frac{4}{5}$.

29. $f(x) = |3x - 4|$; $f(0)$, $f(-5)$, $f(h)$

$f(0) = |3(0) - 4| = |-4| = 4$

$f(-5) = |3(-5) - 4| = |-19| = 19$

$f(h) = |3h - 4|$

So $f(0) = 4$, $f(-5) = 19$, and $f(h) = |3h - 4|$.

C

Identify whether the following relations are functions:

33. $x^2 + y^2 = 16$
$y^2 = 16 - x^2$ Solve for y.
$y = \pm\sqrt{16 - x^2}$

x	y
0	4
0	-4

(0, 4)
(0, -4)

It is not a function.   The value x = 0 is paired with two different values.

37. $|y| + |x| = 3$

x	y
3	0
1	±2

$|y| = -|x| + 3$

$|y| = -|3| + 3 = 0$, so $y = 0$.

$|y| = -|1| + 3 = 2$, so $y = \pm 2$.

It is not a function.   The value x = 1 is paired with two distinct values.

Find the indicated values of $f(x)$:

41. $f(x) = x + 4$; $f(x + h)$, $\dfrac{f(x + h) - f(x)}{h}$

   $f(x + h) = x + h + 4$ Replace x with $x + h$.

   $\begin{aligned}\dfrac{f(x + h) - f(x)}{h} &= \dfrac{x + h + 4 - (x + 4)}{h}\\ &= \dfrac{x + h + 4 - x - 4}{h}\\ &= \dfrac{h}{h}\\ &= 1\end{aligned}$

   So $f(x + h) = x + h + 4$,
   and $\dfrac{f(x + h) - f(x)}{h} = 1$.

45. $f(x) = -3x - 2$; $f(x + h)$, $\dfrac{f(x + h) - f(x)}{h}$

   $f(x + h) = -3(x + h) - 2 = -3x - 3h - 2$

   $\begin{aligned}\dfrac{f(x + h) - f(x)}{h} &= \dfrac{-3x - 3h - 2 - (-3x - 2)}{h}\\ &= \dfrac{-3x - 3h - 2 + 3x + 2}{h}\\ &= \dfrac{-3h}{h}\\ &= -3\end{aligned}$

   So $f(x + h) = -3x - 3h - 2$
   and $\dfrac{f(x + h) - f(x)}{h} = -3$

D

49. The cost of a load of hot asphalt delivered to your business is $45 a cubic yard plus a $50 delivery charge. The cost (c) can be expressed by the equation $c(x) = 45x + 50$. Find the cost of a load of 5 cubic yards of asphalt.

   Formula:
   $c(x) = 45x + 50$
   $c(x) = 45 \cdot 5 + 50$ Replace x by 5.
   $c(x) = 275$

   The cost of a load of 5 cubic yards of asphalt is $275.

E Maintain Your Skills

Solve by any method:

53. $(x - 4)^2 = 4$
$x - 4 = \pm 2$ Find the square root of each side.

$x - 4 = 2$ or $x - 4 = -2$
$x = 6$ $x = 2$

The solution set is $\{2, 6\}$.

57. $m^2 + \frac{2}{5}m = \frac{7}{5}$

$5m^2 + 2m = 7$ Multiply by 5 to clear the fractions.
$5m^2 + 2m - 7 = 0$
$(5m + 7)(m - 1) = 0$
$5m + 7 = 0$ or $m - 1 = 0$
$m = -\frac{7}{5}$ $m = 1$

The solution set is $\left\{-\frac{7}{5}, 1\right\}$.

61. $x^2 + 3x + f = 0$
$a = 1, b = d, c = f$ Use the quadratic formula.

$x = \frac{-d \pm \sqrt{d^2 - 4(1)(f)}}{2(1)}$

$x = \frac{-d \pm \sqrt{d^2 - 4f}}{2}$

229

EXERCISES 6.7   VARIATION

A

1. If x varies directly as y, and if x = 6 when y = 15, find y when x = 30.

   Step 1
   $k = \frac{x}{y}$

   So x = ky

   This is an example of direct variation so the quotient $\frac{x}{y}$ is used.

   Step 2
   6 = 15k

   $k = \frac{6}{15} = \frac{2}{5}$

   Replace x with 6 and y with 15 to find k.

   Step 3
   $30 = \frac{2}{5}y$

   y = 75

   Substitute $k = \frac{2}{5}$ and x = 30 in the formula in Step 1.

5. If x varies inversely as y, and if x = 24 when y = 6, find x when y = 15.

   Step 1
   k = xy

   This is an example of indirect variation, so the product xy is used.

   Step 2
   k = (24)(6) = 144

   Replace x with 24 and y with 6 to find k.

   Step 3
   144 = 15x

   x = 9.6

   Substitute k = 144 and y = 15 in the formula at Step 1.

9. If z varies jointly as x and y, and if z = 36 when x = 4 and y = 9, find z if x = 28 and y = 7.

   Step 1
   $k = \frac{z}{xy}$

   Since this is an example of joint variation, use the quotient $\frac{z}{xy}$.

230

Step 2
$k = \dfrac{36}{4 \cdot 9} = 1$

Replace z with 36, x with 4 and y with 9 and solve for k.

Step 3
$1 = \dfrac{z}{(28)(7)}$

$z = 196$

Substitute x = 28, y = 7, and k = 1 in the formula in Step 1.
Solve for z.

B

13. The pressure P per square inch in water varies directly as the depth d. If P = 5.77 when d = 13, find P when d = 32. (To the nearest tenth.)

Step 1
$k = \dfrac{P}{d}$

This is an example of direct variation so use the quotient, $\dfrac{P}{d}$.

Step 2
$k = \dfrac{5.77}{13}$

Replace P with 5.77 and d with 13.

Step 3
$\dfrac{5.77}{13} = \dfrac{P}{32}$

$P = \dfrac{5.77}{13}(32) \approx 14.2$

Substitute $\dfrac{5.77}{13}$ for k and 32 for d.
Round to the nearest tenth.

So if the distance is 32, the pressure is 14.2.

17. The number of amperes varies directly as the number of watts. For a reading of 550 watts, the number of amperes is 5. What is the number of amperes when the reading of watts is 1600? (To the nearest tenth.)

Step 1
Let a represent amperes, and w represent watts.
$k = \dfrac{w}{a}$

This is an example of direct variation.

Step 2
$k = \dfrac{550}{5} = 110$

To find k, the constant of variation, replace a with 5 and w with 550.

Step 3

$110 = \dfrac{1600}{a}$ Substitute k with 110 and w with 1600.
$110a = 1600$ Solve.
$a = 14.5$ Round to the nearest tenth.

So when the number of watts is 1600, the number of amperes is 14.5.

C

21. At PPL Express the cost of shipping goods (C) varies jointly as the distance shipped (d) and the weight (w). If it costs $338 to ship 6.5 tons of goods 8 miles, how much will it cost to ship 24 tons of goods 1535 miles? (Round to the nearest dollar.)

Step 1
$k = \dfrac{C}{dw}$ This is an example of joint variation.

Step 2
$k = \dfrac{338}{8(6.5)} = 6.5$ Solve for k by replacing C with 338, w with 6.5, and d with 8.

Step 3
$6.5 = \dfrac{C}{24(1535)}$ Substitute d with 1535, w with 24, and k with 6.5.

$C = (6.5)(24)(1535) = 239{,}460$ Solve for C.

It costs $239,460 to ship 24 tons a distance of 1535 miles.

25. The volume of a box with a fixed depth varies jointly as the length and width of the bottom of the box. If the volume of the box is 1152 cm^3 when the dimensions of the bottom are 16 cm × 12 cm, find the volume when the bottom dimensions are 15 cm × 7 cm.

Step 1
Let V represent the volume, ℓ the length and w the width. k represents the constant of variation:

$k = \dfrac{V}{\ell w}$ This is an example of joint variation.

Step 2
$k = \dfrac{1152}{(16)(12)} = 6$ Replace the variables and solve for k.

Step 3

$6 = \dfrac{V}{(15)(7)}$

$V = 630$

Replace the variables and solve for V.

When the dimensions of the bottom of the box are 15 cm × 7 cm, the volume is 630 cm^3.

E Maintain Your Skills

Simplify using only positive exponents:

29. $\left[\dfrac{x^{-1}}{y^2}\right]^{-2} \left[\dfrac{x}{y^{-2}}\right] = \left[\dfrac{x^2}{y^{-4}}\right]\left[\dfrac{x}{y^{-2}}\right] = \dfrac{x^3}{y^{-6}} = x^3 y^6$

Write in scientific notation:

33. $0.0000007843 = 7.843 \times 10^{-7}$ Multiply 7.843 times the −7 power of 10 to represent moving the decimal point 7 places to the right.

CHAPTER 7

SYSTEMS OF LINEAR EQUATIONS AND INEQUALITIES

EXERCISES 7.1 SYSTEMS OF LINEAR EQUATIONS: SOLVING BY GRAPHING AND SUBSTITUTION

A

Solve by graphing:

1. $\begin{cases} x + y = -4 \\ x - y = 8 \end{cases}$

   Write each equation in slope-intercept form to determine the type of system.

   $x + y = -4 \qquad x - y = 8$
   $\quad y = -x - 4 \qquad y = x - 8$

   $m = -1 \qquad\qquad m = 1$
   y-intercept (0, -4)   y-intercept (0, -8)

   The slopes are not the same, so the system has one solution.
   Use the slope and y-intercept of each line to draw the graphs.

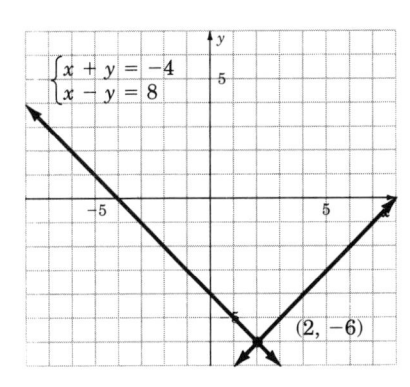

   The solution is {(2, -6)}.

   The lines intersect at (2, -6).

5. $\begin{cases} y = x - 4 \\ y = -x - 2 \end{cases}$

   $y = x - 4 \qquad y = -x - 2$
   $m = 1 \qquad\qquad m = 1$
   y-intercept (0, -4)   y-intercept (0, -2)

   The slopes are not the same, so the system has one solution.
   Draw the graphs.

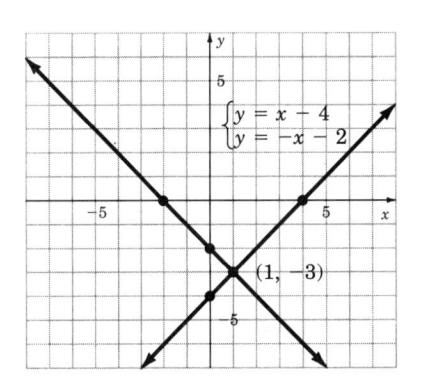

   The solution set is {(1, -3)}.

   The lines intersect at (1, -3).

235

Solve the following systems by substitution:

9. $\begin{cases} y = x + 2 & (1) \\ x + y = 10 & (2) \end{cases}$

$\quad\quad x + y = 10 \quad\quad (2)$ Equation (1) is solved for y,
$\quad x + (x + 2) = 10$ so substitute x + 2 for y in
$\quad\quad\quad\quad 2x + 2 = 10$ equation (2).
$\quad\quad\quad\quad\quad\quad 2x = 8$ Solve for x.
$\quad\quad\quad\quad\quad\quad\quad x = 4$

$\quad\quad y = x + 2 \quad\quad (1)$ To find y, substitute x = 4
$\quad\quad y = 4 + 2$ in equation (1).
$\quad\quad y = 6$ Solve for y.

The solution set is {(4, 6)}.

13. $\begin{cases} x + y = 4 & (1) \\ 3x - 2y = 7 & (2) \end{cases}$

$\quad\quad x + y = 4 \quad\quad (1)$ Solve equation (1) for x.
$\quad\quad x = -y + 4$

$\quad\quad\quad\quad 3x - 2y = 7 \quad\quad (2)$ Substitute -y + 4 for x in
$\quad 3(-y + 4) - 2y = 7$ equation (2).
$\quad\quad -3y + 12 - 2y = 7$
$\quad\quad\quad\quad\quad\quad -5y = -5$
$\quad\quad\quad\quad\quad\quad\quad y = 1$

$\quad\quad x + y = 4 \quad\quad (1)$ Substitute y = 1 in
$\quad\quad x + 1 = 4$ equation (1).
$\quad\quad\quad\quad x = 3$

The solution set is {(3, 1)}.

17. $\begin{cases} y = 2x - 3 & (1) \\ x = 2y - 9 & (2) \end{cases}$

$\quad\quad x = 2y - 9 \quad\quad (2)$ Substitute 2x - 3 for y in
$\quad x = 2(2x - 3) - 9$ equation (2).
$\quad x = 4x - 6 - 9$
$-3x = -15$
$\quad x = 5$

$\quad y = 2x - 3 \quad\quad (1)$ Substitute x = 5 in
$\quad y = 2(5) - 3$ equation (1).
$\quad y = 10 - 3$
$\quad y = 7$

The solution set is {(5, 7)}.

21. $\begin{cases} 4x - y = 16 & (1) \\ 3x + 2y = 1 & (2) \end{cases}$

$\begin{aligned} 4x - y &= 16 \quad (1) \\ 4x - 16 &= y \end{aligned}$ Solve for y in equation (1).

$\begin{aligned} 3x - 2y &= 1 \quad (2) \\ 3x + 2(4x - 16) &= 1 \\ 3x + 8x - 32 &= 1 \\ 11x &= 33 \\ x &= 3 \end{aligned}$ Substitute 4x - 16 for y in equation (2).

$\begin{aligned} 4x - y &= 16 \quad (1) \\ 4(3) - y &= 16 \\ 12 - y &= 16 \\ 12 - 16 &= y \\ -4 &= y \end{aligned}$ Substitute x = 3 in equation (1).

The solution set is {(3, -4)}.

25. $\begin{cases} x + 3y = 5 & (1) \\ 3x - 5y = 12 & (2) \end{cases}$

$\begin{aligned} x + 3y &= 5 \quad (1) \\ x &= -3y + 5 \end{aligned}$ Solve equation (1) for x.

$\begin{aligned} 3x - 5y &= 12 \quad (2) \\ 3(-3y + 5) - 5y &= 12 \\ -9y + 15 - 5y &= 12 \\ -14y &= -3 \\ y &= \frac{3}{14} \end{aligned}$ Substitute -3y + 5 for x in equation (2).

$\begin{aligned} x + 3y &= 5 \quad (1) \\ x + 3\left[\frac{3}{14}\right] &= 5 \\ 14x + 9 &= 70 \\ 14x &= 61 \\ x &= \frac{61}{14} \end{aligned}$ Substitute $\frac{3}{14}$ for y in equation (1).

Solve for x.

The solution set is $\left\{\left[\frac{61}{14}, \frac{3}{14}\right]\right\}$.

C

29. $\begin{cases} x - 3y = 7 & (1) \\ 6x + 4y = 0 & (2) \end{cases}$

$\quad x - 3y = 7 \quad (1)$ Solve for x in equation (1).
$\quad x = 3y + 7$

$\quad 6x + 4y = 9 \quad (2)$ Substitute $3y + 7$ for x in equation (2).
$6(3y + 7) + 4y = 9$
$18y + 42 + 4y = 9$
$\quad\quad\quad 22y = -33$
$\quad\quad\quad\quad y = -\dfrac{33}{22}$
$\quad\quad\quad\quad y = -\dfrac{3}{2}$

$\quad x - 3y = 7 \quad (1)$ Substitute $-\dfrac{3}{2}$ for y in equation (1).
$x - 3\left(-\dfrac{3}{2}\right) = 7$

$2x + 9 = 14$
$\quad 2x = 5$
$\quad\quad x = \dfrac{5}{2}$

The solution set is $\left\{\left(\dfrac{5}{2}, -\dfrac{3}{2}\right)\right\}$.

33. $\begin{cases} x - \dfrac{4}{3}y = -\dfrac{2}{3} & (1) \\ \dfrac{x}{2} + y = \dfrac{5}{2} & (2) \end{cases}$

$x - \dfrac{4}{3}y = -\dfrac{2}{3} \quad (1)$ Solve equation (1) for x.
$\quad\quad x = -\dfrac{2}{3} + \dfrac{4}{3}y$

$\dfrac{x}{2} + y = \dfrac{5}{2} \quad (2)$
$x + 2y = 5 \quad (2^1)$ Multiply equation (2) by 2. Call the result equation (2^1).

$-\dfrac{2}{3} + \dfrac{4}{3}y + 2y = 5$ Substitute $-\dfrac{2}{3} + \dfrac{4}{3}y$ for x in equation (2^1). Solve for y.
$-2 + 4y + 6y = 5$
$\quad\quad\quad 10y = 17$
$\quad\quad\quad\quad y = \dfrac{17}{10}$

$x = -\frac{2}{3} + \frac{4}{3}y$ | Substitute $y = \frac{17}{10}$ in
$x = -\frac{2}{3} + \frac{4}{3} \cdot \frac{17}{10}$ | $x = -\frac{2}{3} + \frac{4}{3}y$ and solve for x.
$x = -\frac{2}{3} + \frac{34}{15}$
$x = -\frac{10}{15} + \frac{34}{15} = \frac{24}{15}$
$x = \frac{8}{5}$

The solution set is $\left\{\left(\frac{8}{5}, \frac{17}{10}\right)\right\}$.

37. $\begin{cases} \frac{1}{4}x + \frac{1}{5}y = -3 & (1) \\ \frac{2}{3}x - \frac{1}{3}y = -13 & (2) \end{cases}$

$\frac{1}{4}x + \frac{1}{5}y = -3$ (1)   Solve equation (1) for x.
$\frac{1}{4}x = -3 - \frac{1}{5}y$
$x = -12 - \frac{4}{5}y$

$\frac{2}{3}x - \frac{1}{3}y = -13$ (2)   Multiply equation (2) by 3, and call the result equation (2^1).

$2x - y = -39$ (2^1)

$2\left(-12 - \frac{4}{5}y\right) - y = -39$ Substitute $x = -12 - \frac{4}{5}y$ in equation (2^1) and solve for y.

$-24 - \frac{8}{5}y - y = -39$
$-120 - 8y - 5y = -195$
$-13y = -75$
$y = \frac{75}{13}$

$x = -12 - \frac{4}{5}y$ Substitute $\frac{75}{13}$ for y in
$x = -12 - \frac{4}{5} \cdot \frac{75}{13}$ $x = -12 - \frac{4}{5}y$ and solve for x.
$x = -12 - \frac{60}{13}$
$x = \frac{-156}{13} - \frac{60}{13}$
$x = -\frac{216}{13}$

The solution set is $\left\{\left(-\frac{216}{13}, \frac{75}{13}\right)\right\}$

D

41. Perry has 27 coins having a total value of $4.95. If these coins are nickels and quarters, how many of each does he have?

Simpler word form:

$\begin{bmatrix} \text{Number} \\ \text{of} \\ \text{nickels} \end{bmatrix} + \begin{bmatrix} \text{Number} \\ \text{of} \\ \text{quarters} \end{bmatrix} = 27$ Write two equations.

$\begin{bmatrix} \text{Value} \\ \text{of} \\ \text{nickels} \end{bmatrix} + \begin{bmatrix} \text{Value} \\ \text{of} \\ \text{quarters} \end{bmatrix} = \$4.95$

Select variables:
Let x represent the number of nickels.
Let y represent the number of quarters.

Translate to algebra:
$\begin{cases} x + y = 27 & (1) \\ 0.05x + 0.25y = 4.95 & (2) \end{cases}$ Each nickel is $.05 and each quarter is $.25.

Solve:

```
x + y = 27                                Solve equation (1) for x.
    x = 27 - y
       0.05x + 0.25y = 4.95               Substitute 27 - y for x in
0.05(27 - y) + 0.25y = 4.95               equation (1).
     1.35 - 0.05y + 0.25y = 4.95
                   0.20y = 3.6
                       y = 18

x + y = 27                                Substitute 18 for y in
x + 18 = 27                               equation (1).
     x = 9
```

The solution set is {(9, 18)}.

Check:
```
    x + y = 27          (1)               Replace x with 9 and y with
   9 + 18 = 27                            18 in each equation.
       27 = 27                            True.

       0.05x + 0.25y = 4.95    (2)
    0.05(9) + 0.25(18) = 4.95
         0.45 + 4.5 = 4.95
                4.95 = 4.95               True.
```

Answer:
Perry has 9 nickels and 18 quarters.

45. The Shocking Electronics Firm makes two kinds of resistors on the same assembly line. One resistor costs $2 to make and takes 3 minutes to assemble. The other one costs $1 to make and takes 2 minutes to assemble. If in one work week (2400 minutes) the cost of production was $1375, how many of each type of resistor was produced?

Simpler word form:

$$\begin{bmatrix} \text{Production} \\ \text{cost of} \\ \text{type 1} \end{bmatrix} + \begin{bmatrix} \text{Production} \\ \text{cost of} \\ \text{type 2} \end{bmatrix} = \$1375$$

$$\begin{bmatrix} \text{Assembly} \\ \text{time of} \\ \text{type 1} \end{bmatrix} + \begin{bmatrix} \text{Assembly} \\ \text{time of} \\ \text{type 2} \end{bmatrix} = 2400 \text{ minutes}$$

Select variables:
Let x represent the number of type one resistors.
Let y represent the number of type two resistors.

Translate to algebra:
$$\begin{cases} 2x + y = 1375 & (1) \\ 3x + 2y = 2400 & (2) \end{cases}$$

Solve:

$y = 1375 - 2x$	Solve equation (1) for y.
$3x + 2(1375 - 2x) = 2400$	Substitute $1375 - 2x$ for y in equation (2).
$3x + 2750 - 4x = 2400$	
$-x = -350$	
$x = 350$	
$2(350) + y = 1375$	Replace 350 for x in equqation (1).
$700 + y = 1375$	
$y = 675$	

The solution set is {(350, 675)}.

Answer:
The Shocking Electronics Firm made 350 resistors at $2 each, and 675 resistors at $1 each.

E Maintain Your Skills

Solve:

49. $(y - 2)^2 - 26(y - 2) + 25 = 0$
$y^2 - 4y + 4 - 26y + 52 + 25 = 0$
$y^2 - 30y + 81 = 0$
$(y - 3)(y - 27) = 0$

$y - 3 = 0$ or $y - 27 = 0$
$y = 3$ or $y = 27$

The solution set is {3, 27}.

Draw the graph of the equation:

53. $2x + 3y = 18$

x	y
0	6
2	$\frac{14}{3}$
-1	$\frac{20}{3}$

Find a minimum of three points by assigning values to x and solving for y.

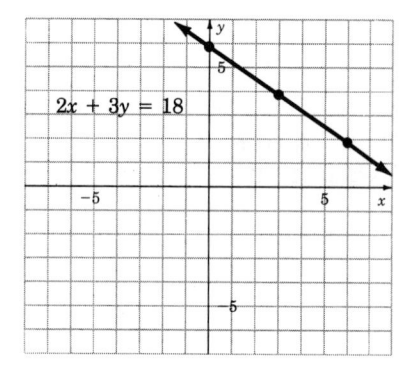

Plot each of the points. Draw the line that passes through the points.

EXERCISES 7.2   SYSTEMS OF LINEAR EQUATIONS (TWO VARIABLES):
                SOLVING BY LINEAR COMBINATIONS

A

Solve using linear combinations:

1. $\begin{cases} x + y = 5 & (1) \\ x - y = 3 & (2) \end{cases}$

$$\begin{aligned} x + y &= 5 \quad (1) \\ \underline{x - y = 3} \quad &(2) \\ 2x &= 8 \\ x &= 4 \end{aligned}$$

Add equations (1) and (2).

Solve for x.

$\begin{aligned} 4 + y &= 5 \\ y &= 1 \end{aligned}$

Substitute x − 4 in equation (1).

The solution set is {(4, 1)}.

242

5. $\begin{cases} 4x - 3y = 8 & (1) \\ -4x + 3y = 7 & (2) \end{cases}$

$\begin{array}{r} 4x - 3y = 8 \\ -4x + 3y = 7 \\ \hline 0 = 15 \end{array}$ Add.

Both variables are eliminated but the result is a contradiction. The system is independent and inconsistent.

There are no solutions.

9. $\begin{cases} x + 3y = 6 & (1) \\ x - y = 2 & (2) \end{cases}$

$\begin{array}{r} -1(x + 3y) = -1 \cdot 6 \\ -x - 3y = -6 \quad (3) \end{array}$ Multiply equation (1) by -1, that the coefficients of the x terms are opposites. Call the result equation (3).

$\begin{array}{r} -x - 3y = -6 \quad (3) \\ x - y = 2 \quad (2) \\ \hline -4y = -4 \\ y = 1 \end{array}$ Solve for y.

$\begin{array}{r} x + 3 \cdot 1 = 6 \\ x = 3 \end{array}$ Substitute $y = 1$ in equation (1).

The solution set is $\{(3, 1)\}$.

B

13. $\begin{cases} x + y = 1 & (1) \\ x - 5y = -23 & (2) \end{cases}$

$\begin{array}{r} -1(x + y) = -1 \cdot 1 \\ -x - y = -1 \quad (3) \end{array}$ Multiply equation (1) by -1 resulting in equation (3).

$\begin{array}{r} -x - y = -1 \\ x - 5y = -23 \\ \hline -6y = -24 \\ y = 4 \end{array}$ Add equations (3) and (2).

$\begin{array}{r} x + 4 = 1 \\ x = -3 \end{array}$ Substitute $y = 4$ in equation (1).

The solution set is $\{(-3, 4)\}$.

17. $\begin{cases} 5x + 4y = 48 & (1) \\ 3x - 7y = 10 & (2) \end{cases}$

$3(5x + 4y) = 3 \cdot 48$
$15x + 12y = 144 \qquad (3)$

Multiply equation (1) by 3 and equation (2) by -5. Call the results equations (3) and (4).

$-5(3x - 7y) = -5 \cdot 10$
$-15x + 35y = -50 \qquad (4)$

$15x + 12y = 144$
$\underline{-15x + 35y = -50}$
$47y = 94$
$y = 2$

Add equations (3) and (4).

$5x + 4 \cdot 2 = 48$
$5x = 40$
$x = 8$

Substitute $y = 2$ in equation (1).

The solution set is $\{(8, 2)\}$.

C

21. $\begin{cases} 4x + 3y = 5 & (1) \\ 12x + 6y = 13 & (2) \end{cases}$

$-3(4x + 3y) = -3 \cdot 5$
$12x - 9y = -15 \qquad (3)$

Multiply equation (1) by -3. Call the results equation (3).

$-12x - 9y = -15$
$\underline{12x + 6y = 13}$
$-3y = -2$
$y = \frac{2}{3}$

Add equations (3) and (2).

$4x + 3\left[\frac{2}{3}\right] = 5$
$4x + 2 = 5$
$4x = 3$
$x = \frac{3}{4}$

Substitute $y = \frac{2}{3}$ in equation (1).

The solution set is $\left\{\left[\frac{3}{4}, \frac{2}{3}\right]\right\}$.

25. $\begin{cases} 5x + 3y = 3 & (1) \\ 7x - 11y = 8 & (2) \end{cases}$

$7(5x + 3y) = 7 \cdot 3$
$35x + 21y = 21 \qquad (3)$ Multiply equation 91) by 7, and call the result equation (3).

$-5(7x - 11y) = -5 \cdot 8$
$-35x + 55y = -40 \qquad (4)$ Multiply equation (2) by -5, and call the result equation (4).

$35x + 21y = 21$
$\underline{-35x + 55y = -40}$ Add equations (3) and (4).
$76y = -19$
$y = \dfrac{-19}{76}$
$y = -\dfrac{1}{4}$

$5x + 3\left[-\dfrac{1}{4}\right] = 3$ Substitute $y = \dfrac{1}{4}$ in
$5x - \dfrac{3}{4} = 3$ equation (1).
$20x - 3 = 12$
$20x = 15$
$x = \dfrac{3}{4}$

The solution set is $\left\{\left[\dfrac{3}{4}, -\dfrac{1}{4}\right]\right\}$.

29. $\begin{cases} 0.6x - 0.7y = 0.53 & (1) \\ 0.5x + 0.2y = 0.05 & (2) \end{cases}$

$60x - 70y = 53 \qquad (3)$
$50x + 20y = 5 \qquad (4)$ Eliminate the decimals and form an equivalent system by multiplying each equation by 100. The result is equations (3) and (4).

$3000x - 3500y = 2650 \qquad (5)$
$\underline{-3000x - 1200y = -300} \qquad (60)$ A third system is formed by multiplying equation (3) by 50 and equation (4) by -60. Add to obtain a linear combination.
$-4700y = 2350$
$y = -0.5$

$0.6x + 0.35 = 0.53$
$0.6x = 0.18$ Substitute $y = -0.5$ in
$x = 0.3$ equation (1). Solve for y.

The solution set is $\{(0.3, -0.5)\}$.

D

33. A 20-pound mixture of shrimp and crabmeat is prepared by the Alaskan Fish Company and is sold for $76.95. If shrimp costs $3.10 a pound and crabmeat costs $4.25 a pound, how many pounds of each were used?

Simpler word form:

$$\begin{bmatrix} \text{Pounds} \\ \text{of} \\ \text{shrimp} \end{bmatrix} + \begin{bmatrix} \text{Pounds} \\ \text{of} \\ \text{crabmeat} \end{bmatrix} = 20 \text{ pounds}$$

$$\begin{bmatrix} \text{Cost} \\ \text{of} \\ \text{shrimp} \end{bmatrix} + \begin{bmatrix} \text{Cost} \\ \text{of} \\ \text{crabmeat} \end{bmatrix} = \$76.95$$

Select variables:
Let x represent the number of pounds of shrimp, and y represent the number of pounds of crabmeat.

Translate to algebra:
$$\begin{cases} x + y = 20 & (1) \\ 3.10x + 4.25y = 76.95 & (2) \end{cases}$$

Solve:

310x + 425y = 7695 (3)	Multiply equation (2) by 100 to obtain an equivalent equation without decimals.
−310x − 310y = −6200	Multiply equation (1) by −310.
115y = 1495	
y = 13	
x + 13 = 20 (1)	
x = 7	

The solution set is {(7, 13)}.

Answer:
In the 20-pound mixture, 7 pounds of shrimp and 13 pounds of crabmeat were used.

37. Statistics from the Girl Scout candy sale show that it takes 5 minutes to sell a $1.75 bar and 4 minutes to sell a $2.25 bar. If a Scout spent 120 minutes selling candy and raised $59.00, how many of each price bar did she sell?

Simpler word form:

$$\begin{bmatrix} \text{Money raised} \\ \text{from} \\ \$1.75 \text{ bars} \end{bmatrix} + \begin{bmatrix} \text{Money raised} \\ \text{from} \\ \$2.25 \text{ bars} \end{bmatrix} = \$59.00$$

$$\begin{bmatrix} \text{Time spent} \\ \text{to sell} \\ \$1.75 \text{ bars} \end{bmatrix} + \begin{bmatrix} \text{Time spent} \\ \text{to sell} \\ \$2.25 \text{ bars} \end{bmatrix} = 120 \text{ minutes}$$

Select variables:
Let x represent the number of $1.75 bars sold, and y represent the number of $2.25 bars.

Translate to algebra:
$$\begin{cases} 1.75x + 2.25y = 59 & (1) \\ 5x + 4y = 120 & (2) \end{cases}$$

Solve:

175x + 225y = 5900 (3)	Multiply equation (1) by 100 to eliminate the decimals.
−175x − 140y = −4200 (4)	Multiply equation (2) by −35.
85y = 1700	Add.
y = 20	
5x + 4(20) = 120	Substitute y = 20 in equation (2).
5x = 40	
x = 8	

The solution set is {(8, 20)}.

Answer:
The Scout sold 8 bars at $1.75 each and 20 bars at $2.25 each.

E   Maintain Your Skills

41. Write the coordinates of the x- and y-intercepts of the graph of 12y = x + 15.

    x − intercept   y-intercept
       y = 0   x = 0
   12·0 = x + 15   12y = 0 + 15
       x = −15 $y = \frac{15}{12} = \frac{5}{4}$

      (−15, 0) $\left(0, \frac{5}{4}\right)$

45. Write the equation, in standard form, of the line containing the points (6, 3) and (9, −5).

    $m = \frac{-5 - 3}{9 - 6} = \frac{-8}{3} = -\frac{8}{3}$ Calculate m.

    $y - 3 = -\frac{8}{3}(x - 6)$ Substitute (6, 3) and m using the point-slope form.

    $y - 3 = -\frac{8}{3}x + 16$

    3y − 9 = −8x + 48
    8x + 3y = 57   Standard form.

EXERCISES 7.3 SYSTEMS OF LINEAR EQUATIONS (THREE VARIABLES):
              SOLVING BY LINEAR COMBINATIONS

A

Solve using linear combinations:

1. $\begin{cases} x + y + z = 4 & (1) \\ x - y + 2z = 8 & (2) \\ 2x + y - z = 3 & (3) \end{cases}$

$\begin{array}{ll} x + y + z = 4 & (1) \\ \underline{x - y + 2z = 8} & (2) \\ 2x \phantom{{}+{}y} + 3z = 12 & \end{array}$ The y term is eliminated if equations (1) and (2) are are added.

$\begin{array}{ll} x - y + 2z = 8 & (2) \\ \underline{2x + y - z = 3} & (3) \\ 3x \phantom{{}+{}y} + z = 11 & (5) \end{array}$ Again eliminate the y term by adding equations (2) and (3).

$\begin{array}{l} -3(3x + z) = -3 \cdot 11 \\ -9x - 3z = -33 \\ \underline{2x + 3z = 12} \\ -7x \phantom{{}+{}3z} = -21 \\ x = 3 \end{array}$ Multiply equation (5) by -3 and add to equation (4) to eliminate z.

$\begin{array}{l} 2 \cdot 3 + 3x = 12 \\ 3z = 6 \\ z = 2 \end{array}$ Substitute x = 3 in equation (4) to find z.

$\begin{array}{l} 3 + y + 2 = 4 \\ y = -1 \end{array}$ Substitute x = 3 and z = 4 in equation (1) to find y.

The solution set is {(3, -1, 2)}.

B

5. $\begin{cases} x + 2y + 3z = 4 & (1) \\ 2x + y + z = 0 & (2) \\ 3x + y + 4z = 2 & (3) \end{cases}$

$\begin{array}{ll} 2x + y + z = 0 & (2) \\ -3x(2x + y + z = -3 \cdot 0) & \\ -6x - 3y - 3z = 0 & \\ \underline{x + 2y + 3z = 4} & (1) \\ -5x - y \phantom{{}+{}3z} = 4 & (4) \end{array}$ Multiply equation (2) by -3 and add to equation (1) to eliminate z.

$\begin{array}{ll} 2x + y + z = 0 & (2) \\ -4(2x + y + z) = -4 \cdot 0 & \\ -8x - 4y - 4z = 0 & \\ \underline{3x + y + 4z = 2} & (3) \\ -5x - 3y \phantom{{}+{}4z} = 2 & (5) \end{array}$ Multiply equation (2) by -4 and add to equation (3) to eliminate z.

248

$$\begin{aligned}
-5x - y &= 4 \\
-1(-5x - y) &= -1 \cdot 4 \\
5x + y &= -4 \\
\underline{-5x - 3y = 2} \quad &(5) \\
-2y &= -2 \\
y &= 1
\end{aligned}$$

Multiply equation (4) by -1 and add to equation (5) to eliminate x.

$$\begin{aligned}
-5x - 1 &= 4 \\
-5x &= 5 \\
x &= -1
\end{aligned}$$

Substitute $y = 1$ in equation (4) to find x.

$$\begin{aligned}
-1 + 2 \cdot 1 + 3z &= 4 \\
3z &= 3 \\
z &= 1
\end{aligned}$$

Substitute $x = -1$ and $y = 1$ in equation (1) to find z.

The solution set is $\{(-1, 1, 1)\}$.

C

9. $\begin{cases} 9x + 3y + 2z = 3 & (1) \\ 4x + 2y + 3z = 9 & (2) \\ 3x + 5y + 4z = 19 & (3) \end{cases}$

$$\begin{aligned}
-27x - 9y - 6z &= -9 \\
\underline{8x + 4y + 6z} &= \underline{18} \\
-19x - 5y &= 9 \quad (4)
\end{aligned}$$

Multiply equation (1) by -3.
Multiply equation (3) by 2.
Add.

$$\begin{aligned}
-16x - 8y - 12z &= -36 \\
\underline{9x + 15y + 12z} &= \underline{57} \\
-7x + 7y &= 21 \quad (5) \\
-x + y &= 3 \quad (6)
\end{aligned}$$

Multiply equation (2) by -4.
Multiply equation (3) by 3.
Add.
Simplify equation (5) by dividing by 7.

$$\begin{aligned}
-5x + 5y &= 15 \\
\underline{-19x - 5y} &= \underline{9} \\
-24x &= 24 \\
x &= -1
\end{aligned}$$

Now multiply equation (6) by 5 and add to equation (5).

$$\begin{aligned}
19 - 5y &= 9 \\
-5y &= -10 \\
y &= 2
\end{aligned}$$

Substitute $x = -1$ in equation (4) to find y.

$$\begin{aligned}
-9 + 6 + 2z &= 3 \\
2z &= 6 \\
z &= 3
\end{aligned}$$

Substitute $x = -1$ and $y = 2$ in equation (1) to find z.

The solution set is $\{(-1, 2, 3)\}$.

13. The general form of a parabola is $y = ax^2 + bx + c$. Find the equation of the parabola that passes through the points $(-2, 0)$, $(-5, 9)$, and $(1, 9)$.

Substitute the values for x and y in the given equation, one pair at a time.

$0 = a(-2)^2 + b(-2) + c$
$0 = 4a - 2b + c$

Substitute $(-2, 0)$ in the equation.

$9 = a(-5)^2 + b(-5) + c$
$9 = 25a - 5b + c$

Substitute $(-5, 9)$.

$9 = a(1)^2 + b(1) + c$
$9 = a + b + c$

Substitute $(1, 9)$.

$\begin{cases} 4a - 2b + c = 0 \quad (1) \\ 25a - 5b + c = 9 \quad (2) \\ a + b + c = 9 \quad (3) \end{cases}$

The result is a system in three variables: a, b, and c.

$\begin{array}{r} -25a + 5b - c = -9 \\ \underline{4a - 2b + c = 0} \quad (1) \\ -21a + 3b = -9 \quad (4) \end{array}$

Multiply equation (2) by -1 and add to equation (1).

$\begin{array}{r} -25a + 5b - c = -9 \\ \underline{a + b + c = 9} \\ -24a + 6b = 0 \quad (5) \end{array}$

Multiply equation (2) by -1 and add to equation (3).

$\begin{array}{r} 42a - 6b = 18 \\ \underline{-24a + 6b = 0} \\ 18a = 18 \\ a = 1 \end{array}$

Multiply equation (4) by -2 and add to equation (5).

$\begin{array}{r} -21 + 3b = -9 \\ 3b = 12 \\ b = 4 \end{array}$

Substitute $a = 1$ in equation (4) and find b.

$4 - 2 \cdot 4 + c = 0$
$c = 4$

Substitute $a = 1$ and $b = 4$ in equation (1), and solve for c.

The solution set is $\{(1, 4, 4)\}$.

Answer:
The equation of the circle is $y = x^2 + 4x + 4$.

Replace a with 1, b with 4, and c with 4 in the given formula.

D

17. The Alpine Fruit Company packs three different gift boxes of pears, apples, and oranges. Box A contains 5 pears, 3 apples, and 1 orange. Box B contains 2 pears, 5 apples, and 2 oranges. Box C contains 3 pears, 4 apples, and 4 oranges. The company has 84 pears, 94 apples, and 50 oranges. How many of each gift box can be prepared if all the fruit is to be used?

Simpler word form:

$$\begin{bmatrix} \text{Pears in} \\ \text{type A} \\ \text{boxes} \end{bmatrix} + \begin{bmatrix} \text{Pears in} \\ \text{type B} \\ \text{boxes} \end{bmatrix} + \begin{bmatrix} \text{Pears in} \\ \text{type C} \\ \text{boxes} \end{bmatrix} = 84$$

$$\begin{bmatrix} \text{Apples in} \\ \text{type A} \\ \text{boxes} \end{bmatrix} + \begin{bmatrix} \text{Apples in} \\ \text{type B} \\ \text{boxes} \end{bmatrix} + \begin{bmatrix} \text{Apples in} \\ \text{type C} \\ \text{boxes} \end{bmatrix} = 94$$

$$\begin{bmatrix} \text{Oranges in} \\ \text{type A} \\ \text{boxes} \end{bmatrix} + \begin{bmatrix} \text{Oranges in} \\ \text{type B} \\ \text{boxes} \end{bmatrix} + \begin{bmatrix} \text{Oranges in} \\ \text{type C} \\ \text{boxes} \end{bmatrix} = 50$$

Select variables:
Let a represent the number of boxes of type A, b represent the number of boxes of type B, and c represent the number of boxes of type C.

Translate to algebra:

$$\begin{cases} 5a + 2b + 3c = 84 & (1) \\ 3a + 5b + 2c = 94 & (2) \\ a + 2b + 4c = 50 & (3) \end{cases}$$

Solve:

$-20a - 8b - 12c = -336$	Multiply equation (1) by -4.
$\underline{9a + 15b + 12c = 282}$	Multiply equation (2) by 3.
$-11a + 7b = -54 \quad (4)$	Add to eliminate the variable c.

$-3a - 5b - 4c = -94$	Multiply equation (2) by -1,
$\underline{a + 2b + 4c = 50}$	and add to equation (3).
$-2a - 3b = -44 \quad (5)$	

$-22a + 14b = -108$	Multiply equation (4) by 2.
$\underline{22a + 33b = 484}$	Multiply equation (5) by -11.
$47b = 387$	Add.
$b = 8$	

$-11a + 5b = -54$ $-11a = -110$ $a = 10$	Substitute b = 8 in equation (4) and solve for a.
$5(10) + 2(8) + 3c = 84$ $50 + 16 + 3c = 84$ $66 + 3c = 84$ $3c = 18$ $c = 6$	Substitute a = 10 and b = 8 in equation (1) and solve for c.

The solution set is {(10, 8, 6)}.

<u>Answer:</u>
The company can prepare 10 gift boxes of type A, 8 of type B, and 6 of type C.

E Maintain Your Skills

Write this equation in standard form:

21. $3x - 18 = y + 7$
     $3x - y = 25$

Write the equation, in standard form, of the line described:

25. The line parallel to y = 2x - 3 and having y-intercept (0, 5).

$y = 2x - 3$	In the line given, m = 2. The given point is the y-intercept, (0, b), so b = 5.
$y = mx + b$ $y = 2x + 5$	Using the slope-intercept form, substitute m = 2, and b = 5.

EXERCISES 7.4   EVALUATING DETERMINANTS

A

Evaluate the following determinants:

1. $\begin{vmatrix} 1 & -1 \\ -1 & 1 \end{vmatrix} = 1 \cdot 1 - (-1 \cdot -1)$

    $= 1 - 1$
    $= 0$

    Find the products indicated by the arrows. Subtract these products.

5. $\begin{vmatrix} 1 & -1 \\ 2 & 3 \end{vmatrix} = 1 \cdot 3 - (2 \cdot -1)$

    $= 3 - (-2)$
    $= 5$

Solve for a:

9. $\begin{vmatrix} a & 1 \\ 1 & 1 \end{vmatrix} = 2$

$\begin{vmatrix} a & 1 \\ 1 & 1 \end{vmatrix} = a \cdot 1 - 1 \cdot 1 = a - 1$ Expand the 2 × 2 determinant and set it equal to 2.

$a - 1 = 2$
$a = 3$

The solution set is {3}.

B

Evaluate the following determinants:

13. $\begin{vmatrix} 4 & 7 \\ 3 & -6 \end{vmatrix} = 4 \cdot -6 - 3 \cdot 7$ Find the products indicated by the arrows and subtract.
$= -24 - 21$
$= -45$

17. $\begin{vmatrix} 1 & 2 & -1 \\ 2 & 1 & 1 \\ 1 & 2 & 2 \end{vmatrix}$ Choose a column or row.

$\begin{vmatrix} 1 & 2 & -1 \\ 2 & 1 & 1 \\ 1 & 2 & 2 \end{vmatrix} = 1(+1) \begin{vmatrix} 1 & 1 \\ 2 & 2 \end{vmatrix} + 2(-1) \begin{vmatrix} 2 & -1 \\ 1 & 2 \end{vmatrix} + 1(+1) \begin{vmatrix} 2 & -1 \\ 1 & 1 \end{vmatrix}$

Using the first columns, expand by minors. From the array of sign the sign of the first minor is, "+", the sign of the second is "-", and the sign of the third is "+".

$= 1[1 \cdot 2 - 1 \cdot 1] - 2[2 \cdot 2 - 1(-1)] + 1[2 \cdot 1 - 1(-1)]$
$= 1(1) - 2(4 + 1) + 1(2 + 1)$
$= 1 - 10 + 3$
$= -6$

21. $\begin{vmatrix} 2 & 3 & 2 \\ 0 & -1 & 0 \\ 5 & 2 & 1 \end{vmatrix}$

$\begin{vmatrix} 2 & 3 & 2 \\ 0 & -1 & 0 \\ 5 & 2 & 1 \end{vmatrix} = -0 \begin{vmatrix} 3 & 2 \\ 2 & 1 \end{vmatrix} + (-1) \begin{vmatrix} 2 & 2 \\ 5 & 1 \end{vmatrix} - 0 \begin{vmatrix} 2 & 3 \\ 5 & 2 \end{vmatrix}$ Use the second row, and expand by minors.

$= 0 - 1 \cdot 2 \cdot 1 - 5 \cdot 2) = 0$
$= -1(2 - 10)$
$= 8$

C

Evaluate the following determinants:

25. $\begin{vmatrix} -3 & -3 & 1 \\ 4 & 1 & 4 \\ 2 & 2 & -3 \end{vmatrix}$

$\begin{vmatrix} -3 & -3 & 1 \\ 4 & 1 & 4 \\ 2 & 2 & -3 \end{vmatrix} = -4 \begin{vmatrix} -3 & 1 \\ 2 & -3 \end{vmatrix} + 1 \begin{vmatrix} -3 & 1 \\ 2 & -3 \end{vmatrix} - 4 \begin{vmatrix} -3 & -3 \\ 2 & 2 \end{vmatrix}$ Use the second row and expand by minors.

$ = -4(9 - 2) + (9 - 2) - 4[-6 - (-6)]$ The signs of the minors alternate: "−", "+", "−".
$ = -4(7) + 7 - 4(0)$
$ = -28 + 7$
$ = -21$

29. $\begin{vmatrix} 0 & 2 & -5 \\ 1 & 3 & 6 \\ 0 & 4 & 1 \end{vmatrix} = 0 \begin{vmatrix} 3 & 6 \\ 4 & 1 \end{vmatrix} - 1 \begin{vmatrix} 2 & -5 \\ 4 & 1 \end{vmatrix} + 0 \begin{vmatrix} 2 & -5 \\ 3 & 6 \end{vmatrix}$ Use the first column and expand by minors.

$ = 0 - 1[2 - (-20)] + 0$
$ = -1(22)$
$ = -22$

Solve for a:

33. $\begin{vmatrix} -1 & -2 & 1 \\ a & 0 & 0 \\ -1 & 1 & 1 \end{vmatrix} = 3$

$\begin{vmatrix} -1 & -2 & 1 \\ a & 0 & 0 \\ -1 & 1 & 1 \end{vmatrix} = -1 \begin{vmatrix} -2 & 1 \\ 1 & 1 \end{vmatrix} + 0 \begin{vmatrix} -1 & 1 \\ -1 & 1 \end{vmatrix} - 0 \begin{vmatrix} -1 & -2 \\ -1 & 1 \end{vmatrix}$ Expand about the second row.

$ = -a(-2 - 1) + 0 - 0$
$ = -a(-3)$
$ = 3a$
$3a = 3$ Set 3a equal to 3 and solve for a.
$a = 1$

The solution set is {1}.

E Maintain Your Skills

Solve:

37. $\dfrac{1}{x} - \dfrac{1}{x+3} = 2$

  $x + 3 - x = 2x(x + 3)$ Multiply by the LCM: $x(x+3)$.
  $3 = 2x^2 + 6x$
  $2x^2 + 6x - 3$
  $a = 2, b = 6, c = -3$ Use the quadratic formula.

  $x = \dfrac{-6 \pm \sqrt{6^2 - 4(-6)}}{4}$

  $x = \dfrac{-6 \pm \sqrt{60}}{4}$

  $x = \dfrac{-6 \pm 2\sqrt{15}}{4}$

  $x = \dfrac{2(-3 \pm \sqrt{15})}{2 \cdot 2}$ Factor and reduce.

  $x = \dfrac{-3 \pm \sqrt{15}}{2}$

41. $\sqrt{x + 11} - 3 = \sqrt{x - 4}$

  $x + 11 - 6\sqrt{x + 11} + 9 = x - 4$ Square both sides.
  $-6\sqrt{x + 11} = -24$
  $\sqrt{x + 11} = 4$ Divide both sides by -6.
  $x + 11 = 16$ Square both sides again.
  $x = 5$

45. $|2x - 1| < 11$

  So $2x - 1 = 11$ or $2x - 1 = -11$ Find critical numbers by
  $2x = 12$ or $2x = -10$ setting both sides equal.
  $x = 6$ or $x = -5$

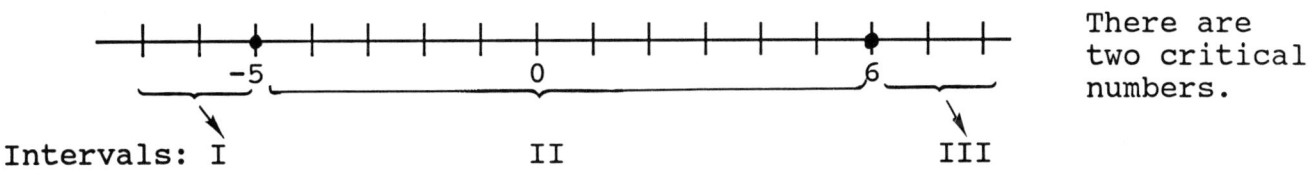

  There are two critical numbers.

  Intervals: I   II   III

I	II	III
$x = -7$	$x = 0$	$x = 10$
$|2 \cdot -7 - 1| < 11$	$|-1| < 11$	$|2 \cdot 10 - 1| < 11$
$|-15| < 11$	$1 < 11$	$|19| < 11$
False	True	False

  Test a value from each interval defined by the critical numbers.

  So the solution set is $\{x | -5 < x < 6\}$ or, in interval notation: $(-5, 6)$.

EXERCISES 7.5 SYSTEMS OF LINEAR EQUATIONS: SOLVING BY DETERMINANTS

A

Solving using Cramer's Rule:

1. $\begin{cases} 2x + 5y = 15 \\ 3x - 4y = 11 \end{cases}$

$D = \begin{vmatrix} 2 & 5 \\ 3 & -4 \end{vmatrix}$, $D_x = \begin{vmatrix} 15 & 5 \\ 11 & -4 \end{vmatrix}$, $D_y = \begin{vmatrix} 2 & 15 \\ 3 & 11 \end{vmatrix}$ Write the determinants D, D_x, and D_y. For D_x replace the first column of D with the constant terms. For D_y replace the second column of D with the constant terms.

$D = 2(-4) - 3(5) = -8 - 15 = -23$
$D_x = 15(-4) - 11(5) = -60 - 55 = -115$
$D_y = 2(11) - 3(15) = 22 - 45 = -23$

$x = \dfrac{D_x}{D} = \dfrac{-115}{-23} = 5$ Use Cramer's rule to solve for x and y.

$y = \dfrac{D_y}{D} = \dfrac{-23}{-23} = 1$

The solution set is $\{(5, 1)\}$.

5. $\begin{cases} 3x + 4y = 10 \\ 2x - 3y = 1 \end{cases}$

$D = \begin{vmatrix} 3 & 4 \\ 2 & -3 \end{vmatrix}$, $D_x = \begin{vmatrix} 10 & 4 \\ 1 & -3 \end{vmatrix}$, $D_y = \begin{vmatrix} 3 & 10 \\ 2 & 1 \end{vmatrix}$ Write the determinants. D, D_x, D_y. Expand D, D_x, D_y.

$D = 3(-3) - 2(4) = -9 - 8 = -17$
$D_x = 10(-3) - 1(4) = -30 - 4 = -34$
$D_y = 3(1) - 2(10) = 3 - 20 = -17$

$x = \dfrac{D_x}{D} = \dfrac{-34}{-17} = 2$ Solve for x and y.

$y = \dfrac{D_y}{D} = \dfrac{-17}{-17} = 1$

The solution set is $\{(2, 1)\}$.

B

9. $\begin{cases} x - 3y = 4 \\ 3x - 9y = 5 \end{cases}$

$D = \begin{vmatrix} 1 & -3 \\ 3 & -9 \end{vmatrix}, \; D_x = \begin{vmatrix} 4 & -3 \\ 5 & -9 \end{vmatrix}, \; D_y = \begin{vmatrix} 1 & 4 \\ 3 & 5 \end{vmatrix}$

$D = 1(-9) - 3(-3) = -9 + 9 = 0$ $D = 0$; therefore, there is no unique solution.

The system has no unique solution.

C

13. $\begin{aligned} 2x + 3y + z &= 6 \\ x - 2y + 3z &= -3 \\ 3x + y - z &= 8 \end{aligned}$

Write the determinants D, D_x, D_y, and D_z. To form D_x, replace the first column of D with the constant terms. To form D_y, replace the second column D with the constant terms. To form D_z, replace the third column with the constant terms.

$D = \begin{vmatrix} 2 & 3 & 1 \\ 1 & -2 & 3 \\ 3 & 1 & -1 \end{vmatrix} = 2(-1)-1(-4)+3(11)=35$ Expand about the first column. The value of each minor is in parentheses.

$D_x = \begin{vmatrix} 6 & 3 & 1 \\ -3 & -2 & 3 \\ 8 & 1 & -1 \end{vmatrix} = 6(-1)-3(-21)+1(13)=70$ Expand about the first row.

$D_y = \begin{vmatrix} 2 & 6 & 1 \\ 1 & -3 & 3 \\ 3 & 8 & -1 \end{vmatrix} = 2(-21)-1(-14)+3(21)=35$ Expand about the first column.

$D_z = \begin{vmatrix} 2 & 3 & 6 \\ 1 & -2 & -3 \\ 3 & 1 & 8 \end{vmatrix} = 2(-13)-1(18)+3(3) = -35$ Expand about the first column.

$x = \dfrac{D_x}{D} = \dfrac{70}{35} = 2$

$y = \dfrac{D_y}{D} = \dfrac{34}{35} = 1$

$z = \dfrac{D_z}{D} = \dfrac{-35}{35} = -1$

Use Cramer's rule to solve for x, y, and z.

The solution set is $\{(2, 1, -1)\}$.

D

17. The sum of two numbers is 1459. The second number is 85 more than the first. What are the two numbers?

Simpler word form:
A first number + a second number = 1459.
The second number = the first number + 85.

Select variables:
Let x represent the first number and y represent the second number.

Translate to algebra:
$$\begin{cases} x + y = 1459 & (1) \\ y = x + 85 & (2) \end{cases}$$

Solve:
$x + y = 1459$ (1)
$-x + y = 85$ (2)   Rewrite the second equation.

$D = \begin{vmatrix} 1 & 1 \\ -1 & 1 \end{vmatrix} = 1 - (-1) = 2$ Write the determinants D, D_x, and D_y.

$D_x = \begin{vmatrix} 1459 & 1 \\ 85 & 1 \end{vmatrix} = 1459 - 85 = 1374$

$D_y = \begin{vmatrix} 1 & 1459 \\ -1 & 85 \end{vmatrix} = 85 - (-1459) = 15444$

$x = \dfrac{D_x}{D} = \dfrac{1374}{2} = 687$ Use Cramer's rule to solve for x and y.

$y = \dfrac{D_y}{D} = \dfrac{1544}{2} = 772$

The solution set is {(687, 772)}.

Answer:
The numbers are 687 and 772.

21. A stock broker buys two stocks, one at $37.50 a share and the other at $14.75 a share. If the total purchase consisted of 650 shares at a cost of $14,137.50, how many shares were purchased at each price?

Simpler word form:
$\begin{bmatrix} \text{Number of} \\ \$37.50 \\ \text{shares} \end{bmatrix} + \begin{bmatrix} \text{Number of} \\ \$14.75 \\ \text{shares} \end{bmatrix} = 650$ shares

$\begin{bmatrix} \text{Cost of} \\ \$37.50 \\ \text{shares} \end{bmatrix} + \begin{bmatrix} \text{Cost of} \\ \$14.75 \\ \text{shares} \end{bmatrix} = \$14{,}137.50$

Select variables:
Let x represent the number of shares purchased at $37.50 a share, and y represent the number of shares at $14.75 a share.

Translate to algebra:
$$x + y = 650 \quad (1)$$
$$37.50x + 14.75y = 14137.50 \quad (2)$$

Solve:
$$x + y = 650 \quad (1)$$
$$3750x + 1475y = 1413750 \quad (2)$$

Multiply equation (2) by 100 to eliminate the decimals.

$$D = \begin{vmatrix} 1 & 1 \\ 3750 & 1475 \end{vmatrix} = -2275$$

Write the determinants D, D_x, and D_y.

$$D_x = \begin{vmatrix} 650 & 1 \\ 1413750 & 1475 \end{vmatrix} = -455000$$

$$D_y = \begin{vmatrix} 1 & 650 \\ 3750 & 1413750 \end{vmatrix} = -1023750$$

$$x = \frac{D_x}{D} = \frac{-455000}{-2275} = 200$$

$$y = \frac{D_y}{D} = \frac{-1023750}{-2275} = 450$$

Use Cramer's rule to solve for x and y.

The solution set is {(200, 450)}.

Answer:
At $37.50 a share, 200 shares were purchased, and 450 shares were purchased at $14.75 a share.

E Maintain Your Skills

Solve:

25. $\sqrt{2x - 3} - \sqrt{x + 10} = -1$

$\sqrt{2x - 3} = -1 + \sqrt{x + 10}$

$2x - 3 = 1 - 2\sqrt{x + 10} + x + 10$ Square both sides.

$x - 14 = -2\sqrt{x + 10}$

$x^2 - 28x + 196 = 4(x + 10)$ Square both sides again.

$x^2 - 28x + 196 = 4x + 40$

$x^2 - 32x + 156 = 0$

$(x - 26)(x - 6) = 0$ Factor.

$x = 26$ or $x = 6$

<u>Check:</u>

$\sqrt{2 \cdot 26 - 3} - \sqrt{26 + 10} = -1$ Substitute $x = 26$ in original equation.

$\sqrt{49} - \sqrt{36} = -1$

$7 - 6 = -1$

$1 = -1$ False. So reject $x = 26$.

$\sqrt{2 \cdot 6 - 3} - \sqrt{6 + 10} = -1$ Substitute $x = 6$ in original equation.

$\sqrt{9} - \sqrt{16} = -1$

$3 - 4 = -1$

$-1 = -1$ True.

The solution set is $\{6\}$.

29. $(x + 5)^2 - (2x - 7)^2 = 3x^2 + 8$

$x^2 + 10x + 25 - (4x^2 - 28x + 49) = 3x^2 + 8$

$x^2 + 10x + 25 - 4x^2 + 28x - 49 = 3x^2 + 8$

$-3x^2 + 38x - 24 = 3x^2 + 8$

$6x^2 - 38x + 32 = 0$

$3x^2 - 19x + 16 = 0$

$(3x - 16)(x - 1) = 0$

$x = \frac{16}{3}$ or $x = 1$

The solution set is $\left\{\frac{16}{3}, 1\right\}$.

Draw the graph of the inequality:

33. $y > -\frac{3}{2}x = 4$

$y = -\frac{3}{2}x - 4$

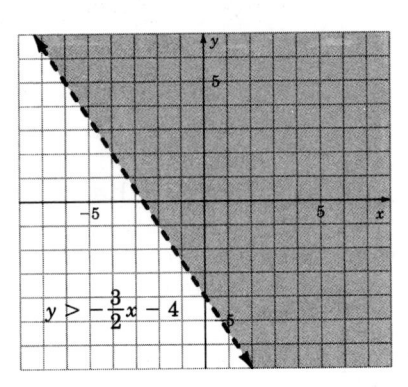

Draw the graph of the corresponding equality.

x	y
0	-4
-2	-1
-4	2

The graph of the line is broken since the equality is not included.
Test the point (0, 0).

$0 > -\frac{3}{2} \cdot 0 - 4$

$0 > -4$ True.

The graph is the half-plane containing the origin.

EXERCISES 7.6 SYSTEMS OF LINEAR INEQUALITIES: SOLVING BY GRAPHING

A

Solve the following systems by graphing:

1. $\begin{cases} x + y \leq 1 \\ x - y \leq 1 \end{cases}$

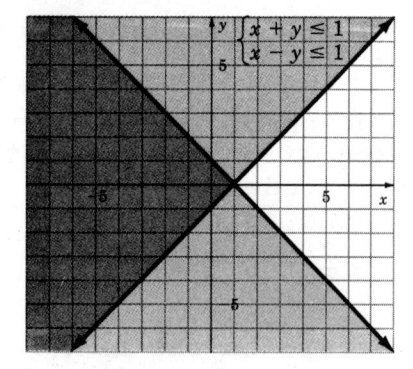

The graph of the solution is the double-shaded area.

To graph $x + y \leq 1$, first graph $x + y = 1$. This graph is a solid line, since equality is part of the relation. Test the point (0, 0) to determine which closed half-plane makes the inequality true. Since (0, 0) makes the inequality true, the closed half-plane containing (0, 0) is the graph of $x + y \leq 1$.

To graph $x - y \leq 1$, first graph $x - y = 1$. This graph is also a solid line. Use (0, 0) as a test point. It makes the inequality true, so the closed half-plane containing (0, 0) is the graph of $x - y \leq 1$.

5. $\begin{cases} 3x + 2y \leq 12 \\ 2x + 5y \geq 10 \end{cases}$

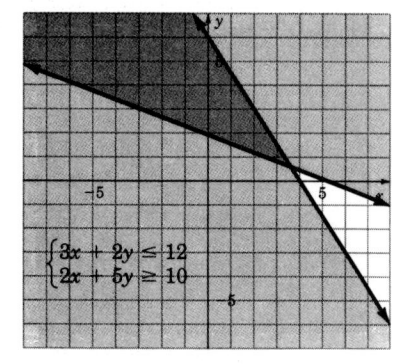

The graph of the solution is the double-shaded area.

To graph $3x + 2y \leq 12$, first graph $3x + 2y = 12$. Use a solid line. Test the point (0, 0). It makes the inequality true, so the closed half-plane containing the origin is the graph of $3x + 2y \leq 12$. To graph $2x + 5y \geq 10$, first graph the equality $2x + 5y = 10$ with a solid line. Then test (0, 0). It makes the inequality false, so the closed half-lane not containing the origin is the graph of $2x + 5y \geq 10$.

B

9. $\begin{cases} 5x + 2y \leq 12 \\ 4x + 5y \geq 10 \end{cases}$

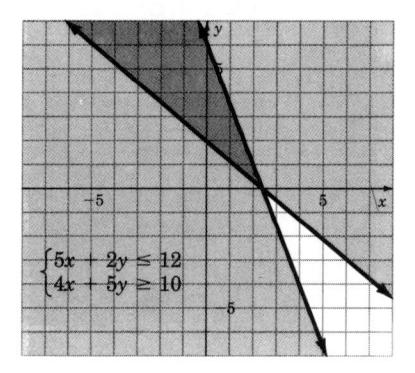

First graph $5x + 2y = 12$. Then use $(0, 0)$ as a test point in the inequality. The closed half-plane containing the origin is the graph of $5x + 2y \leq 12$. Now graph $4x + 5y \geq 10$. Again use a solid line. Using $(0, 0)$ as a test point, determine that the closed half-plane not containing the origin is the graph of $4x + 5y \geq 10$.

C

13. $\begin{cases} \frac{1}{3}x - \frac{1}{4}y \leq 1 \\ \frac{1}{3}x + \frac{1}{4}y \geq 1 \end{cases}$

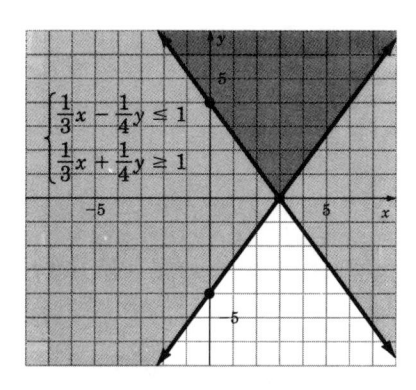

Graph $\frac{1}{3}x - \frac{1}{4}y = 1$. Use the origin as a test point. The closed half-plane containing the origin is the graph of $\frac{1}{3}x - \frac{1}{4}y \leq 1$. Graph $\frac{1}{3}x + \frac{1}{4}y \geq 1$. Use the origin as a test point. The closed half-plane not containing the origin is the graph of $\frac{1}{3}x + \frac{1}{4}y \geq 1$.

The graph of the solution is the double-shaded area.

D

17. At most, $10,000 is to be invested in funds paying 15% and 16%, respectively. If the return from the investments must be at least $1500, what amount of money can be invested at each rate?

Simpler word form:

$\begin{bmatrix} \text{Amount} \\ \text{invested} \\ \text{at 15\%} \end{bmatrix} + \begin{bmatrix} \text{Amount} \\ \text{invested} \\ \text{at 16\%} \end{bmatrix} \leq \$10000$

$\begin{bmatrix} \text{Interest} \\ \text{at} \\ 15\% \end{bmatrix} + \begin{bmatrix} \text{Interest} \\ \text{at} \\ 16\% \end{bmatrix} \geq \$1500$

Select variables:
Let x represent the amount invested at 15% and y represent the amount invested at 16%.

	15%	16%	TOTAL
Money invested	x	y	x + y
Interest earned	0.15x	0.16y	0.15x + 0.16y

Translate to algebra:
$$x + y \leq 10000$$
$$0.15x + 0.16y \geq 1500$$

Solve:

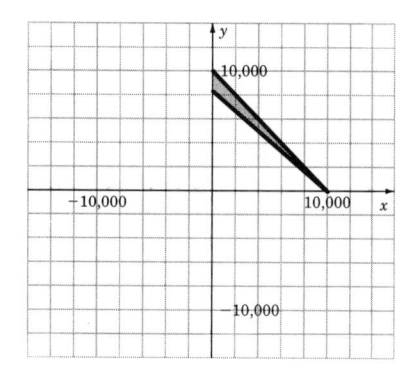

Each unit on the x and y axes is 2000.

Only positive values for x and y need to be considered, since the amount of money invested cannot be negative. To graph x + y ≤ 10000, first graph x + y = 10000. Use a solid line, since equality is part of the relation. Shade the positive values of the appropriate closed half-plane. Then graph 0.15x + 0.16y = 1500. Use a solid line. Shade the positive values of the appropriate closed half-plane.

Answwer:
The points in the double-shaded area together with the points on the lines x + y = 10000 and and 0.15x + 0.16y = 1500 are solutions.

Some examples of solutions are (0, 10000), (4000, 6000), (9000, 950), and (1500, 8000).

E   Maintain Your Skills

21. Write the relation $\{(x, y) | y^2 = 5x - 1, x = 2, 4, 10, 15\}$ as a set of ordered pairs.

x	$y^2 = 5x - 1$		
2	$y^2 = 5 \cdot 2 - 1 = 9$,	$y = \pm 3$	(2, 3) (2, -3)
4	$y^2 = 5 \cdot 4 - 1 = 19$,	$y = \pm\sqrt{19}$	(4, $\sqrt{19}$) (4, $-\sqrt{19}$)
10	$y^2 = 5 \cdot 10 - 1 = 49$,	$y = \pm 7$	(10, 7) (10, -7)
15	$y^2 = 5 \cdot 15 - 1 = 74$,	$y = \pm\sqrt{74}$	(15, $\sqrt{74}$) (15, $-\sqrt{74}$)

The relation as a set of ordered pairs is $\{(2, \pm 3), (4, \pm\sqrt{19}), (10, \pm 7), (15, \pm\sqrt{74})\}$

State the domain and range of the relation:

25. $\{(x, y) | y = |x + 6| - 2\}$

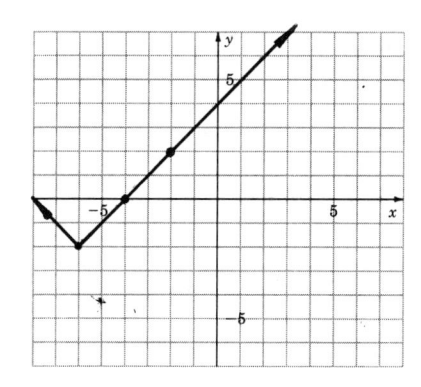

Draw the graph. Make a table of values.

x	y
4	8
-2	2
-4	0
-10	2
-8	0
-6	-2

Domain: $\{x | x \in \mathbb{R}\}$ or $\mathbb{R}$

Range: $\{y | y \geq -2, y \in \mathbb{R}$

29. $\{(x, y) | y^2 = 4x - 6\}$

Is this relation a function?

x	y
2	$\pm\sqrt{2}$
3	$\pm\sqrt{6}$
5	$\pm 2\sqrt{6}$

List some ordered pairs.

It is not a function.

For each of the values of x, there are two different y-values.

CHAPTER 8

SECOND-DEGREE EQUATIONS IN TWO VARIABLES

EXERCISES 8.1 PARABOLAS: QUADRATIC FUNCTIONS

A

Find the vertex, determine whether the vertex is a maximum or minimum, find the x- and y-intercepts, determine the equation of the axis of symmetry, and graph.

1. $f(x) = x^2 - 2$
   $a = 1, b = 0, c = -2$

   a) $h = -\dfrac{0}{2 \cdot 1} = -$

   $f(0) = -2$

   The x-coordinate of the vertex (h) is found by letting $h = -\dfrac{6}{2a}$.

   Determine the y-coordinate of the vertex by substituting 0 for x in the equation.

   The vertex is at (0, -2).

   b) The vertex is a minimum point. The curve opens upward.

   $a > 0$.

   c) $0 = x^2 - 2$
   $2 = x^2$
   $\pm\sqrt{2} = x$

   Substitute 0 for $f(x)$ and solve for x to find the x-intercepts.

   The x-intercepts are $(-\sqrt{2}, 0)$ and $(\sqrt{2}, 0)$.

   $f(0) = -2$

   Substitute 0 for x to find the y-intercept.

   The y-intercept is (0, -2).

   d) Since the vertex is at (0, -2), $x = 0$ is the equation of the axis of symmetry.

e)

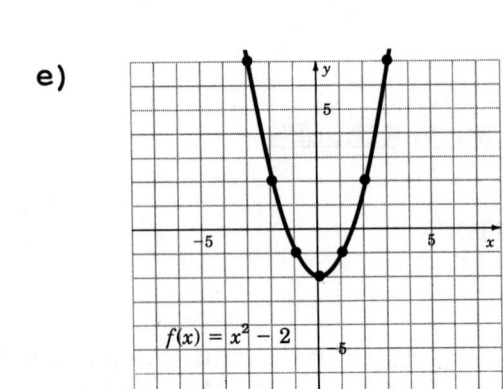

Plot the graph. Three points are known: (0, -2), $(-\sqrt{2}, 0)$, and $(\sqrt{2}, 0)$. Find two other points such as (-3, 7) and (3, 7).

5. $y = (x + 1)^2 - 2$
    $a = 1$, $h = -1$, and $k = -2$ Identify a, h, and k in $f(x) = a(x - h)^2 + k$.

    a) The vertex is $(-1, -2)$ Identify the vertex (h, k).

    b) The vertex is a minimum point. The curve opens upward. $a > 0$.

    c) $0 = (x + 1)^2 - 2$
    $0 = x^2 + 2x + 1 - 2$
    $0 = x^2 + 2x - 1$ Substitute 0 for y and solve for x to find the x-intercepts.

    $x = \dfrac{-2 \pm \sqrt{4 - 4(1)(-1)}}{2}$ Quadratic formula.

    $= \dfrac{-2 \pm 2\sqrt{2}}{2}$

    $= -1 \pm \sqrt{2}$

    The x-intercepts are $(-1 - \sqrt{2}, 0)$ and $-1 + \sqrt{2}, 0)$.

    $y = (0 + 1)^2 - 2$
    $= 1 - 2$
    $= -1$ Substitute 0 for x to find the y-intercept.

    The y-intercept is $(0, -1)$.

    d) Since the vertex is at $(-1, -2)$, $x = -1$ is the equation of the axis of symmetry.

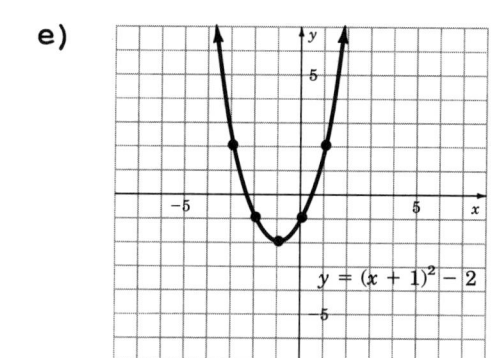

    e)   Plot the graph. Three points are known: $(-1, -2)$, $(-1 -\sqrt{2}, 0)$, and $(-1 +\sqrt{2}, 0)$. Find two more such as $(2, 7)$ and $(-3, 2)$.

9. $y = -x^2 + 3$
   $a = -1$, $b = 0$, $c = 3$

   a) $h = -\dfrac{0}{2(-1)} = 0$

   The x-coordinate of the vertex (h) is found by letting $h = -\dfrac{b}{2a}$.

   $y = -0^2 + 3 = 3$

   Substitute 0 for x to find the y-coordinate of the vertex.

   The vertex is at (0, 3).

   b) The vertex is a maximum point. The curve opens downward.

   $a < 0$.

   c) $0 = -x^2 + 3$
   $x^2 = 3$
   $x = \pm\sqrt{3}$

   Substitute 0 for y and solve for x to find the x-intercepts.

   The x-intercepts are $(-\sqrt{3}, 0)$ and $(\sqrt{3}, 0)$.

   $y = -0^2 + 3$
   $y = 3$

   Substitute 0 for x and solve to find the y-intercept.

   The y-intercept is (0, 3).

   d) Since the vertex is at (0, 3), $x = 0$ is the equation of the axis of symmetry.

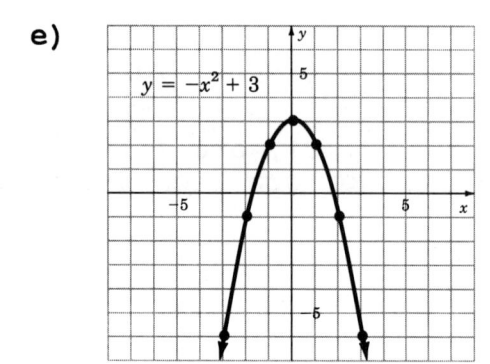

   e)

   Plot the graph. Three points are known: (0, 3), $(-\sqrt{3}, 0)$, and $(\sqrt{3}, 0)$. Find two more, such as (-2, -1) and (2, -1).

270

B

13. $y = -\frac{1}{3}(x - 3)^2 + 3$

$a = -\frac{1}{3}$, $y = 3$, and $k = 3$ — Identify a, h, and k in $f(x) = a(x - h)^2 + k$.

a) The vertex is at (3, 3). — Identify the vertex, (h, k).

b) The vertex is a maximum point, and the curve opens downward. — $a < 0$.

c)
$$0 = -\frac{1}{3}(x - 3)^2 + 3$$
$$0 = -(x - 3)^2 + 9$$
$$x^2 - 6x + 9 = 9$$
$$x^2 - 6x = 0$$
$$x(x - 6) = 0$$
$$x = 0 \text{ or } x = 6$$

Substitute 0 for y and solve for x.

The x-intercepts are (0, 0) and (6, 0).

$$y = -\frac{1}{3}(0 - 3)^2 + 3$$
$$= -\frac{1}{3}(9) + 3$$
$$= 0$$

Substitute 0 for x and solve for y.

The y-intercept is (0, 0).

d) Since the vertex is at (3, 3), $x = 3$ is the equation of the axis of symmetry.

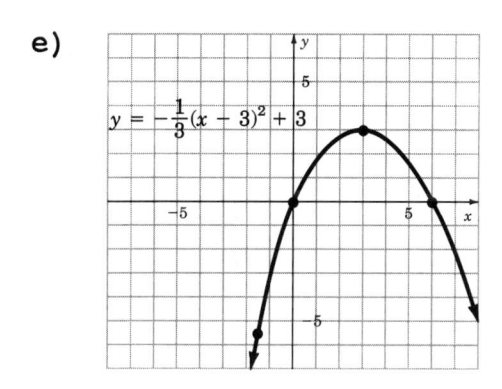

e) Plot the graph. Three points are known: (3, 3), (0, 0), and (6, 0). Find two more, such as $\left[4, \frac{8}{3}\right]$ and by symmetry $\left[2, \frac{8}{3}\right]$.

271

17. $y = \frac{1}{4}(x + 1)^2 - 3$

    $a = \frac{1}{4}$, $h = -1$, $k = -3$ Identify a, h, and k.

   a) The vertex is at $(-1, -3)$.   The vertex is at (h, k).

   b) The vertex is a minimum point $a > 0$.
       and the curve opens upward.

   c) $0 = \frac{1}{4}(x + 1)^2 - 3$ Substitute 0 for y.
       $0 = (x + 1)^2 = 12$
       $0 = x^2 + 2x + 1 - 12$
       $0 = x^2 + 2x - 11$

       $x = \frac{-2 \pm \sqrt{4 - 4(1)(-11)}}{2}$ Use the quadratic formla
                                               and solve for x.
        $= \frac{-2 \pm 4\sqrt{3}}{2}$

        $= -1 \pm 2\sqrt{3}$

      The x-intercepts are $(-1 - 2\sqrt{3}, 0)$
      and $(-1 + 2\sqrt{3}, 0)$.

      $y = \frac{1}{4}(0 + 1)^2 - 3$ Substitute 0 for x.
      $y = \frac{1}{4} - 3 = -\frac{11}{4}$

      The y-intercept is $\left(0, -\frac{11}{4}\right)$.

   d) Since the vertex is at $(-1, -3)$,
      $x = -1$ is the equation of the
      axis of symmetry.

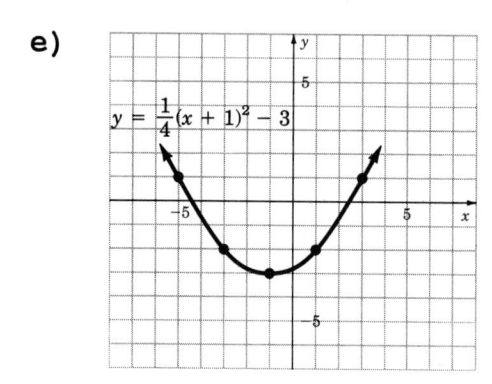

   e) Plot the graph. Three points are known: $(-1, -3)$, $(-1 - 2\sqrt{3}, 0)$, and $(-1 + 2\sqrt{3}, 0)$. Find two more points, such as $(1, -2)$ and by symmetry $(-3, -2)$.

21. $y = x^2 - 4x + 7$
    $a = 1$, $b = -4$, and $c = 7$

a) $h = -\dfrac{-4}{2} = 2$
    $f(2) = 2^2 - 4 \cdot 2 + 7$
    $= 3$

    The vertex is at (2, 3).

    Let $h = -\dfrac{b}{2a}$, where h is the x-coordinate of the vertex. Then evaluate f(h) to find the y-coordinate of the vertex.

b) The vertex is a minimum point. The curve opens upward.

    $a > 0$.

c) $0 = x^2 - 4x + 7$
    $x = \dfrac{4 \pm \sqrt{16 - 4(1)(7)}}{2}$
    $x = \dfrac{4 \pm \sqrt{-12}}{2}$

    Substitute 0 for y and solve for x. Use the quadratic formula.

    There is no x-intercept since $\sqrt{-12}$ is not a real number.

    Since there is no x-intercept and the parabola has a minimum point, the entire curve is above the x-axis.

    $y = 0^2 - 4 \cdot 0 + 7$
    $= 7$

    Substitute 0 for x and solve for y.

    The y-intercept is (0, 7).

d) Since the vertex is at (2, 3), the equation of the axis of symmetry is $x = 2$.

e)
    Plot the graph. Two points are known: (2, 3) and (0, 7). By symmetry: (4, 7). Find two other points, such as (1, 4), and (3, 4).

C

25. $y = 2x^2 + 4x - 2$
    $a = 2$, $b = 4$, and $c = -2$

   a) $h = -\dfrac{4}{2 \cdot 2} = -1$ Find (h, k). $h = -\dfrac{b}{2a}$ and
      $f(-1) = 2(-1)^2 + 4(-1) - 2$ $k = f(h)$.
      $ = 2 - 4 - 2 = -4$

      The vertex is at $(-1, -4)$.

   b) The vertex is a minimum $a > 0$.
      point and the curve opens
      upward.

   c) $0 = 2x^2 + 4x - 2$ Substitute 0 for y, and use
      $a = 2$, $b = 4$, $c = -2$ the quadratic formula to
      $x = \dfrac{-4 \pm \sqrt{16 - 4(2)(-2)}}{2 \cdot 2}$ solve for x.
      $ = \dfrac{-4 \pm 4\sqrt{2}}{4}$
      $ = -1 \pm \sqrt{2}$

      The x-intercepts are $(-1-\sqrt{2}, 0)$
      and $(-1+\sqrt{2}, 0)$.

      $y = 2(0)^2 + 4(0) - 2$ Substitute 0 for x, and solve
      $ = -2$ for y.

      The y-intercept is $(0, -2)$.

   d) Since the vertex is at $(-1, -4)$,
      the axis of symmetry is $x = -1$.

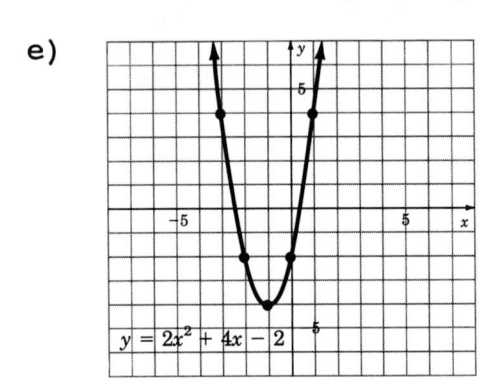

   e) [graph of $y = 2x^2 + 4x - 2$]   Plot the graph using
                                                                 $(-1, -4)$, $(-1-\sqrt{2}, 0)$,
                                                                 $(-1+\sqrt{2}, 0)$ and $(0, -2)$. By
                                                                 symmetry: $(-2, -2)$.

274

29. $y = -2x^2 + 12x - 19$
    $a = -2$, $b = 12$, $c = -19$

   a) $h = -\dfrac{12}{2(-2)} = 3$
       $f(3) = -2(3)^2 + 12 \cdot 3 - 19 = -1$

       The vertex is $(3, -1)$.

   b) The vertex is a maximum point since $a < 0$. The curve opens downward.

   c) $0 = -2x^2 + 12x - 19$ Substitute 0 for y. Solve for x.
      $x = \dfrac{-12 \pm \sqrt{144 - 4(-2)(-19)}}{2(-2)}$
      $= \dfrac{-12 \pm \sqrt{-8}}{-4}$

      Since $\sqrt{-8}$ is a complex number, there are no x-intercepts.

      $y = -2(0)^2 + 12(0) - 19$ Substitute 0 for x.
      $= -19$

      The y-intercept is $(0, -19)$.

   d) Since the vertex is $(3, -1)$, the axis of symmetry is $x = 3$.

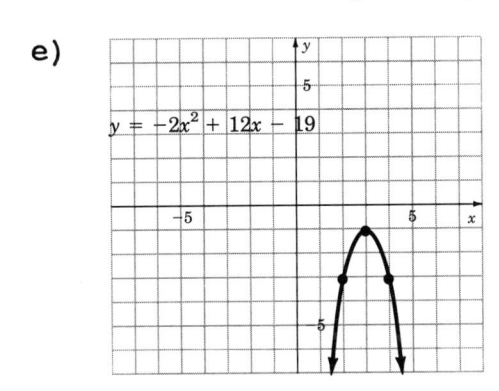

   e) Plot the graph. Only two points are known: $(3, -1)$ and $(0, -19)$. Find $(6, -19)$ by symmetry. Then find two other points, such as: $(2, -3)$ and $(4, -3)$.

D

33. The number of units (u) that a company can produce with n employees is given by $u(n) = -\frac{1}{3}n^2 + 12n$, with $0 < n \leq 30$. Find the maximum number of units that can be produced and the number of employees needed to produce them.

$u(n) = -\frac{1}{3}n^2 + 12n$

Since $a = -\frac{1}{3}$, the graph has a maximum point. To find the maximum point, write the equation in the form:
$f(x) = a(x - h)^2 + k$.

$= -\frac{1}{3}(n^2 - 36n)$

$= -\frac{1}{3}(n^2 - 36n + 324) + 108$

$= -\frac{1}{3}(n - 18)^2 + 108$

$a = -\frac{1}{3}$, $h = 18$, and $k = 108$

Since $h = 18$ and $k = 108$, the vertex is at (18, 108).

Answer:
The number of units that can be made is 108, and the number of employees needed to produce them is 18.

E   Maintain Your Skills

37. If $f(x) = -5x^2 + 7x - 2$, find $f(-2)$, $f(1)$, and $f(5)$.

$f(x) = -5x^2 + 7x - 2$
$f(-2) = -5(-2)^2 + 7(-2) - 2 = -36$ Substitute -2 for x.
$f(1) = -5(1)^2 + 7(1) - 2 = 0$ Substitute 1 for x.
$f(5) = -5(5)^2 + 7(5) - 2 = -92$ Substitute 5 for x.

So $f(-2) = -36$, $f(1) = 0$, and $f(5) = -92$.

Draw the graph of each equation:

41. $4x + 5y = 20$
       $5y = -4x + 20$ Write in slope-intercept
       $y = -\frac{4}{5}x + 4$ form.

   $m = -\frac{4}{5}$ and the y-intercept is (0, 4).

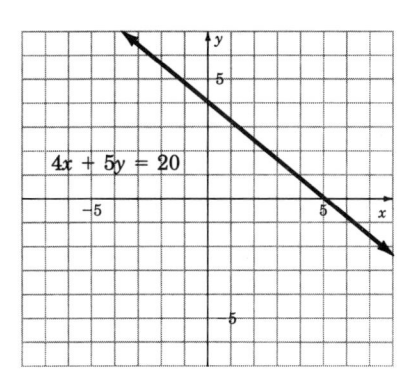

   Plot the y-intercept. Move 5 units right and 4 units down.

45. $y = \sqrt{9 - x^2}$

x	y
0	3
±3	0
±15	≈2.6

   List some ordered pairs.

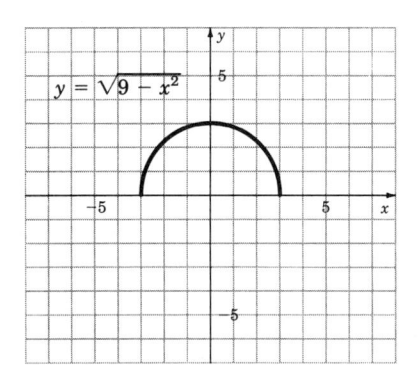

   Plot the points.
   Note that $-3 \leq x \leq 3$ and $0 \leq y \leq 3$.

277

EXERCISES 8.2   CIRCLES

A

Write the coordinates of the center and the length of the radius and graph the following:

1. $x^2 + y^2 = 25$
    Center (0, 0)

    Radius 5

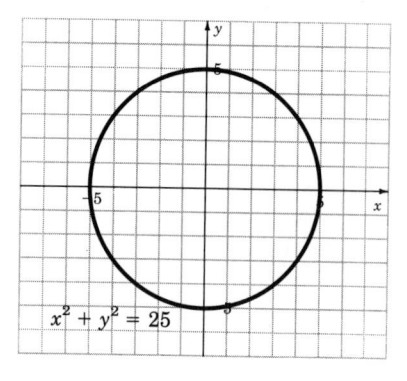

    This is the standard form of an equation whose graph is a circle with center at the origin.
    The radius is the square root of the constant term.

5. $2x^2 + 2y^2 = 8$
    $x^2 + y^2 = 4$

    Center (0, 0)
    Radius 2

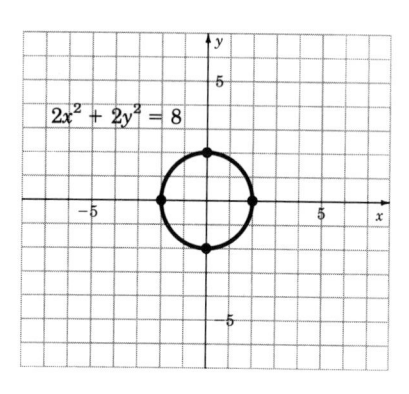

    Divide both sides by the common factor, 2.
    From the standard form.
    The radius is the square root of the constant term.

Write the equation in standard form of the circle with the given coordinates as center and the given value of r as radius:

9. (2, 4), r = 7
    $(x - h)^2 + (y - k)^2 = r^2$

    $(x - 2)^2 + (y - 4)^2 = 49$

    Standard form of a circle with center at (h, k) and radius, r.
    Substitute.

13. $(1, 1), r = 6$
    $(x - 1)^2 + (y - 1)^2 = 36$

    Substitute 1 for h, 1 for k, and 6 for r is the standard form of a circle with center at (h, k) and radius, r.

B

Write the coordinates of the center and the length of the radius and graph each of the following:

17. $(x + 5)^2 + (y - 3)^2 = 4$
    $[x - (-5)]^2 + (y - 3)^2 = 2^2$

    Center (-5, 3)
    Radius 2

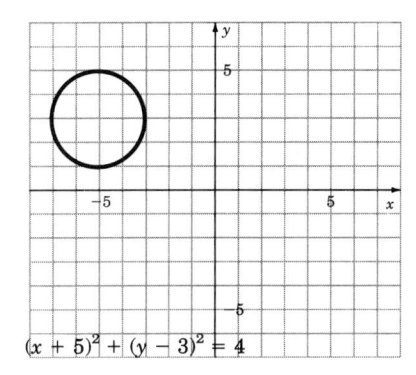

    From the standard form:
    h = -5 and k = 3.
    Radius = $\sqrt{4}$.

21. $(x - 1)^2 + (y - 2)^2 = \frac{9}{4}$

    Center (1, 2)
    Radius $\frac{3}{2}$

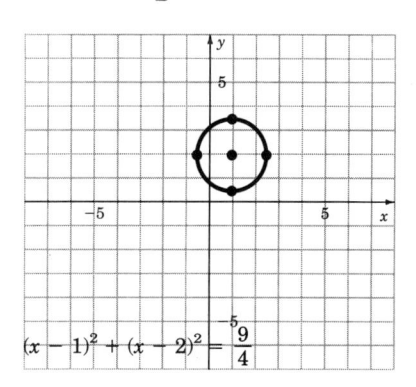

    From the standard form.
    $\sqrt{\frac{9}{4}} = \frac{3}{2}$

279

Write the equation in standard form of the circle with the given coordinates as center and the given value of r as radius:

25. $(-3, -8)$, $r = 16$

$(x - h)^2 + (y - k)^2 = r^2$ Standard form of the equation of a circle with center at (h, k) and radius r.

$[x - (-3)]^2 + [y - (-8)]^2 = (16)^2$ Substitute -3 for h, -8 for k and 16 for r.

$(x + 3)^2 + (y + 8)^2 = 256$

C

Write the coordinates of the center and the length of the radius of the following:

29. $4x^2 + 4y^2 = 9$

$x^2 + y^2 = \frac{9}{4}$ Divide by 4 on both sides. Standard form of a circle with center at $(0, 0)$.

Center $(0, 0)$

Radius $\frac{3}{2}$

33. $x^2 + y^2 - 4x - 6y + 9 = 0$

$x^2 - 4x + y^2 - 6y = -9$

$x^2 - 4x + 4 + y^2 - 6y + 9 = -9 + 4 + 9$ Complete the square in both x and y.

$(x - 2)^2 + (y - 3)^2 = 4$

Center $(2, 3)$

Radius 2 From standard form: $h = 2$, $k = 3$, and $r = 2$.

Write the equation in standard form of the circle with the given coordinates as the center and the given value of r as the radius:

37. $(2, 2)$, $r = \sqrt{2}$

$(x - h)^2 + (y - k)^2 = r^2$ Standard form.

$(x - 2)^2 + (y - 2)^2 = 2$ Substitute 2 for h, 2 for k, and $\sqrt{2}$ for r.

Each of the following is an equation whose graph is a circle. Write each in standard form, then write the coordinates of the center and the length of the radius:

41. $x^2 + y^2 - 10x + 4y + 17 = 0$

$x^2 - 10x + y^2 + 4y = -17$

$x^2 - 10x + 25 + y^2 + 4y + 4 = -17 + 25 + 4$ Complete the square in both x and y.

$(x - 5)^2 + (y + 2)^2 = 12$

Center $(5, -2)$

Radius $2\sqrt{3}$ $h = 5$, $k = -2$, and $r = 2\sqrt{3}$ (or $\sqrt{12}$).

45. $x^2 + y^2 - x + 3y = \frac{3}{2}$

$x^2 - x \qquad + y^2 + 3y \qquad = \frac{3}{2}$ Complete the square in both x and y.

$x^2 - x + \frac{1}{4} + y^2 + 3y + \frac{9}{4} = \frac{3}{2} + \frac{1}{4} + \frac{9}{4}$

$\left[x - \frac{1}{2}\right]^2 + \left[y + \frac{3}{2}\right]^2 = 4$ $h = \frac{1}{2},\ k = -\frac{3}{2},\ r = 2.$

Center $\left[\frac{1}{2},\ -\frac{3}{2}\right]$

Radius 2

Given the center and radius of a circle, write an equation in the form of $ax^2 + by^2 + cx + dy = e = 0$.

49. $(11, -7),\ r = \sqrt{5}$

$(x - h)^2 + (y - k)^2 = r^2$ First, write the standard-

$(x - 11)^2 + [y - (-7)]^2 = (\sqrt{5})^2$ form equation given $h = 11$,

$(x = 11)^2 + (y + 7)^2 = 5$ $h = 11,\ k = -7$ and $r = \sqrt{5}.$

$x^2 - 22x + 121 + y^2 + 14y + 49 = 5$

$x^2 + y^2 - 22x + 14y + 165 = 0$

E   Maintain Your Skills

Solve:

53. $4\sqrt{x - 1} + 2 = 4x - 1$

$4\sqrt{x - 1} = 4x - 3$

$16(x - 1) = (4x - 3)^2$ Square both sides.

$16x - 16 = 16x^2 - 24x + 9$

$16x^2 - 40x + 25 = 0$

$(4x - 5)(4x - 5) = 0$

$x = \frac{5}{4}$ A double root.

Check:

$4\sqrt{\frac{5}{4} - 1} + 2 = 4\left[\frac{5}{4}\right] - 1$

$4\sqrt{\frac{1}{4}} + 2 = 5 - 1$

$4\left[\frac{1}{2}\right] + 2 = 5 - 1$

$2 + 2 = 4$

$4 = 4$ True.

The solution set is $\left\{\frac{5}{4}\right\}$.

57. $7x^2 - 11x + 4 = 0$

$$x = \frac{11 \pm \sqrt{121 - 112}}{14}$$

$$x = \frac{11 \pm \sqrt{9}}{14}$$

$$x = \frac{11 \pm 3}{14}$$

$$x = \frac{11 + 3}{14} \text{ or } x = \frac{11 - 3}{14}$$

$$x = 1 \text{ or } x = \frac{8}{14} = \frac{4}{7}$$

The solution set is $\left\{\frac{4}{7}, 1\right\}$.

61. $\frac{3}{x - 7} < 0$

$x \neq 7$ Restrict x.

The critical number is 7.

```
   3      + +  |  + +
 x - 7    - -  |  + +        The numerator is always
───────────────○──────        positive.
               7
```

The solution set is $(-\infty, 7)$.

EXERCISES 8.3   ELLIPSES AND HYPERBOLAS

A

Draw the graphs of the following conic sections:

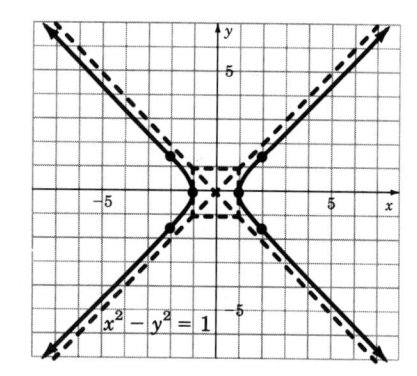

1. $x^2 - y^2 = 1$ Original equation.

$\frac{x^2}{1} - \frac{y^2}{1} = 1$ Write in standard form. This is the standard form of a hyperbola with intercepts at $(1, 0)$ and $(-1, 0)$.

Table of values.

x	y
±1	0
±√2	±1
±2	±√3

5. $\dfrac{x^2}{4} - \dfrac{y^2}{4} = 1$

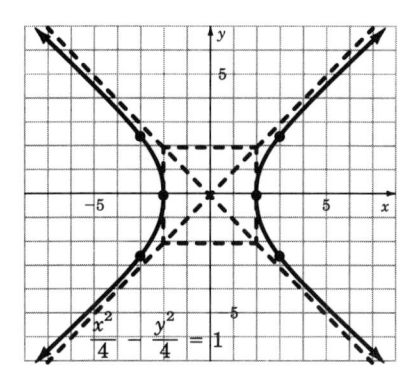

Standard form, with intercepts at (2, 0) and (-2, 0).

Table of values:

x	y
±2	0
±4	±2√3
±6	±4√2

B

9. $\dfrac{x^2}{9} + \dfrac{y^2}{25} = 1$

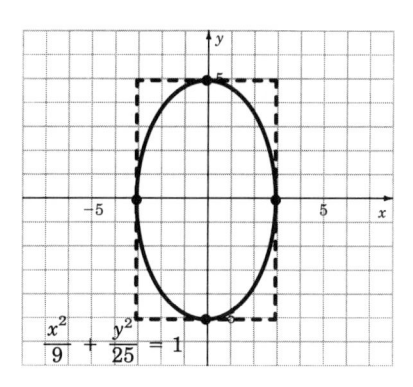

Standard form of an ellipse with intercepts at (±3, 0) and (0, ±5).

Table of values:

x	y
±3	0
0	±5
±2	$\pm\dfrac{5\sqrt{5}}{3} \approx \pm 3.7$
±1	$\pm\dfrac{10\sqrt{2}}{3} \approx \pm 4.8$

13. $\dfrac{x^2}{9} - \dfrac{y^2}{25} = 1$

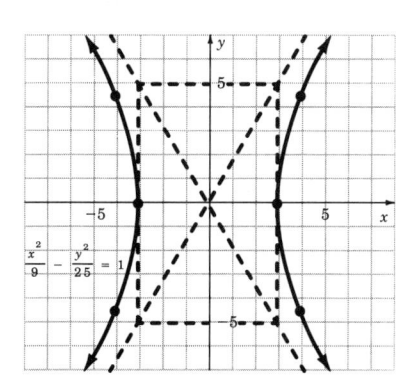

Standard form of a hyperbola with intercepts at (±3, 0).

Table of values:

x	y
±3	0
±5	$\pm\dfrac{20}{3} \approx \pm 6.7$
±4	$\pm\dfrac{5\sqrt{7}}{3} \approx \pm 4.4$

C

17. $\dfrac{x^2}{27} + \dfrac{y^2}{3} = 3$

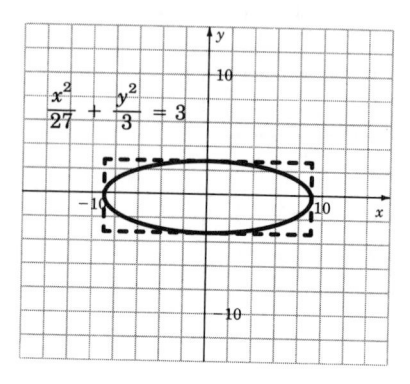

Standard form of an ellipse with intercepts at $(\pm 3\sqrt{3}, 0)$ and $(0, \pm\sqrt{3})$.

Table of values:

x	y
$\pm 3\sqrt{3} \approx \pm 5.2$	0
0	$\pm\sqrt{3} \approx \pm 1.7$
± 1	$\pm\dfrac{4\sqrt{5}}{3} \approx \pm 3.0$
± 3	$\pm 2\sqrt{2} \approx \pm 2.8$

21. $5x^2 - 6y^2 = 60$

$\dfrac{x^2}{12} - \dfrac{y^2}{10} = 1$

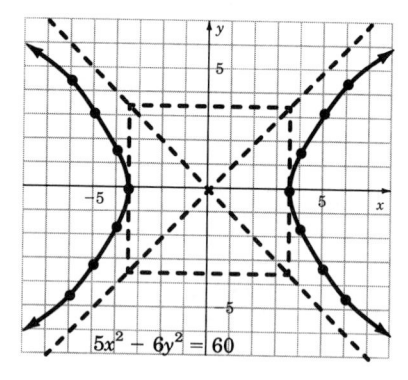

Original equation.

Write in standard form. Divide both sides by 60. This is the standard form of a hyperbola with intercepts at $(\pm 2\sqrt{3}, 0)$.

Table of values:

x	y
$\pm 2\sqrt{3} \approx \pm 3.5$	0
± 4	$\approx \pm 1.8$
± 5	$\approx \pm 3.3$
± 6	$\approx \pm 4.5$

Write the equation of each ellipse of hyperbola in standard form:

25.

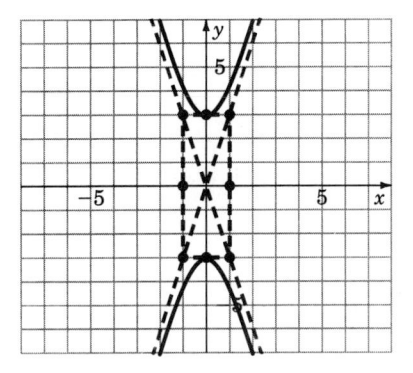

The y-intercepts are (0, 3) and (0, -3).

The x-intercepts of the rectangle are (1, 0) and (-1, 0).

$$\frac{y^2}{9} - \frac{x^2}{1} = 1$$

From the graph write the coordinates of the y-intercepts. The y-intercepts are (0, ±b). Write the x-coordinates of the rectangle. The x-coordinates are (±a, 0). Substitute the values of a and b in the standard form $\frac{y^2}{b^2} - \frac{x^2}{a^2} = 1$.

29.

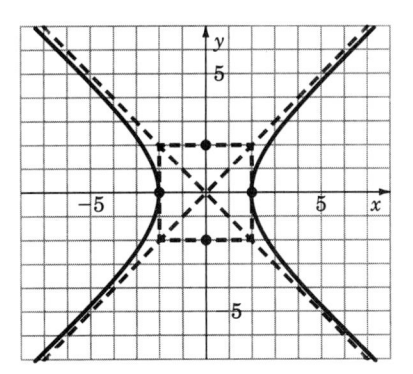

The x-intercepts are (2, 0) and (-2, 0).

The y-intercepts of the rectangle are (0, 2) and (0, -2).

$$\frac{x^2}{4} - \frac{y^2}{4} = 1$$

From the graph, write the coordinates of the x-intercepts of the hyperbola. The x-intercepts are (±a, 0). Then write the y-intercepts of the rectangle. The y-intercepts of the rectangle are at (0, ±b). Substitute the values of a and b in the standard form: $\frac{x^2}{a^2} - \frac{y^2}{b^2} = 1$.

285

E Maintain Your Skills

Solve:

33. $\dfrac{1}{a} + \dfrac{1}{a-1} = \dfrac{2}{a^2 + 2a - 3}$

$\dfrac{1}{a} + \dfrac{1}{a-1} = \dfrac{2}{(a+3)(a-1)}$

$(a-1)(a+3) + a(a+3) = 2(a)$ Multiply both sides by the LCM of the denominators: $a(1-1)(a+3)$.

$a^2 + 2a - 3 + a^2 + 3a = 2a$
$2a^2 + 3a - 3 = 0$

$a = \dfrac{-3 \pm \sqrt{9 + 24}}{4}$ Use the quadratic formula to solve for a.

$a = \dfrac{-3 \pm \sqrt{33}}{4}$

Solve for w:

37. $3w^2 - 8cw - 3c^2 = 0$
$a = 3,\ b = -8c,\ c = -3c^2$ Use the quadratic formlua.

$w = \dfrac{8c + \sqrt{64c^2 + 36c^2}}{6}$

$w = \dfrac{8c + 10c}{6}$

$w = \dfrac{8c + 10c}{6}$ or $w = \dfrac{8c - 10c}{6}$

$w = \dfrac{18c}{6} = 3c$ or $w = \dfrac{-2c}{6} = -\dfrac{c}{3}$

So $w = 3c$ or $w = -\dfrac{c}{3}$.

Solve by graphing:

41. $\begin{cases} y = x + 3 \\ x = 2y + 1 \end{cases}$

$y = x + 3$ $x = 2y + 1$ Write each equation in slope-
                         $2y = x - 1$ intercept form. Find the
                         $y = \frac{x}{2} - \frac{1}{2}$ slope and the y-intercept of
                                                   each.

$m = 1$ $m = \frac{1}{2}$

y-intercept $(0, 3)$ y-intercept $\left(0, -\frac{1}{2}\right)$

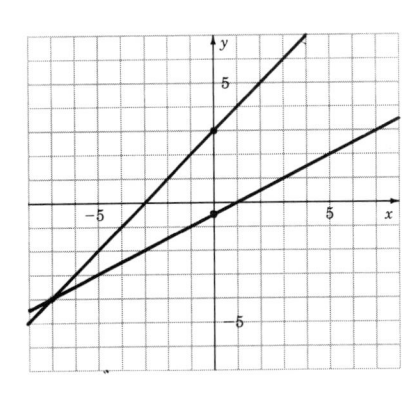

Graph both lines using their y-intercepts and slopes.

The solution set is $\{(-7, -4)\}$.

EXERCISES 8.4   INVERSE OF A RELATION

A

Each of the following equations defines a function R. Write the equation defining F^{-1} and draw the graphs of F and F^{-1} on the same axes. State whether F^{-1} is a function:

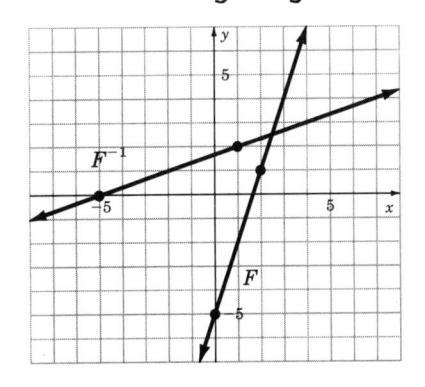

1. $y = 3x - 5$

   $F: \quad y = 3x - 5$
   $F^{-1}: \quad x = 3y - 5$ To write the rule for the
   $\qquad 3y = x + 5$ inverse function, interchange
   $\qquad y = \frac{x}{3} + \frac{5}{3}$ the variables in the rule for
                                           F. Solve for y.

F^{-1} is a function.   From the graph.

287

5. $-\frac{1}{2}x + \frac{1}{3}y = 1$
    $-3x + 2y = 6$
    $2y = 3x + 6$ Solve for y.
    $y = \frac{3}{2}x + 3$ Standard form.

    F: $y = \frac{3}{2}x + 3$

    F^{-1}: $x = \frac{3}{2}y + 3$ To write the rule for the inverse function, interchange the variables in the rule for the function.
    $2x = 3y + 6$
    $2x - 6 = 3y$
    $\frac{2}{3}x - 2 = y$ Solve for y.

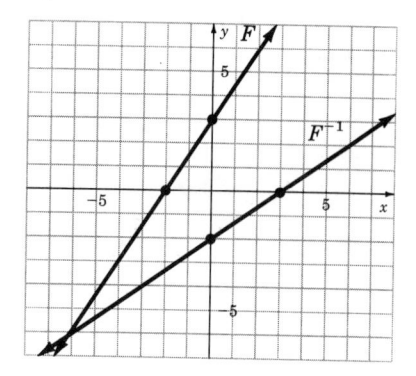

    F^{-1} is a function.   From the graph.

9. $y = \frac{1}{3}x^2$

    F: $y = \frac{1}{3}x^2$

    F^{-1}: $x = \frac{1}{3}y^2$ To write the rule for F^{-1}, interchange the variables in the rule for F and solve for y.
    $3x = y^2$
    $\pm\sqrt{3x} = y$

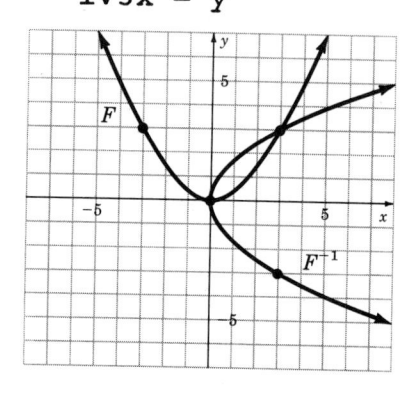

    F^{-1} is not a function.   From the graph. So F is a function, but not a one-to-one function.

288

B

13. $y = 2(x - 3)^2 + 1$
   F: $y = 2(x - 3)^2 + 1$
   F^{-1}: $x = 2(y - 3)^2 + 1$
   $x - 1 = 2(y - 3)^2$
   $\dfrac{x - 1}{2} = (y - 3)^2$
   $\pm\sqrt{\dfrac{x - 1}{2}} = y - 3$
   $3 \pm \sqrt{\dfrac{x - 1}{2}} = y$

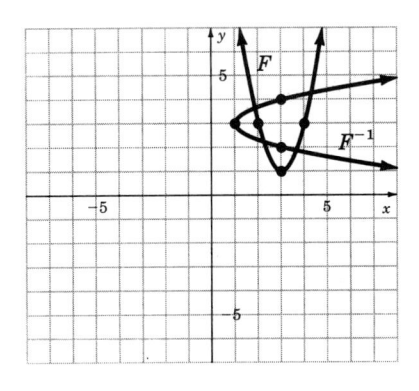

F^{-1} is not a function.   From the graph.

17. $y = 2$
   F: $y = 0x + 2$
   F^{-1}: $x = 0y + 2$
   $x = 2$

Standard form.

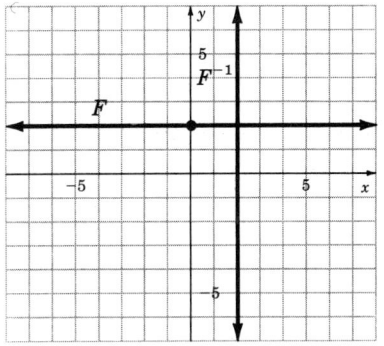

F^{-1} is not a function.

289

C

21. $y = \sqrt{3x - 6}$

    F: $y = \sqrt{3x - 6}$ with $y \geq 0$.

    F^{-1}: $x = \sqrt{3y - 6}$, $x \geq 0$.

$$x^2 = 3y - 6$$
$$x^2 + 6 = 3y$$
$$\frac{x^2 + 6}{3} = y \text{ with } x \geq 0.$$

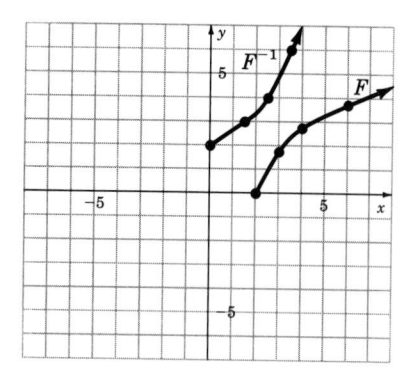

Note that $y \geq 0$, since the principal root is always non-negative.

Note that $x \geq 0$ since the variables are interchanged.

F^{-1} is a function.

25. $y = \sqrt{x - 5} + 3$

    F: $y = \sqrt{x - 5} + 3$, $y \geq 0$

    F^{-1}: $x = \sqrt{y - 5} + 3$, $x \geq 0$

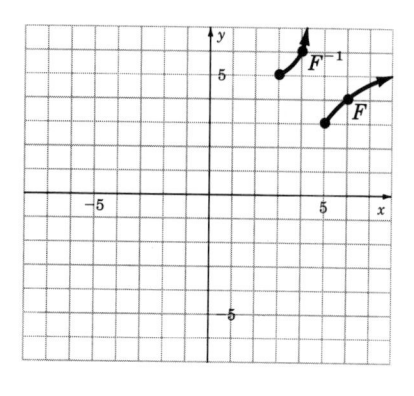

F^{-1} is a function.

29. $y = |x| + 1$
    F: $y = |x| + 1$, $x \in \mathbb{R}$, $y \geq 1$

    F^{-1}: $x = |y| + 1$, $x \geq 1$, $y \in \mathbb{R}$

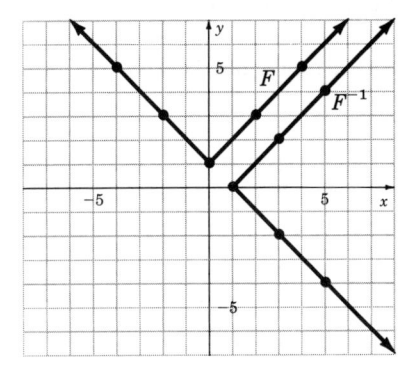

F^{-1} is not a function.

E Maintain Your Skills

33. Find the slope of the line through the points (6, -5) and (-3, -2).

$$m = \frac{y_2 - y_1}{x_2 - x_1}$$

$$m = \frac{-2 - (-5)}{-3 - 6} = \frac{-2 + 5}{-9} = -\frac{3}{9} = -\frac{1}{3}$$

The slope of the line is $-\frac{1}{3}$.

37. Find the equations of both lines that are perpendicular to the line segment joining (2, 5) and (4, -6) and that pass through the endpoints of the segment.

$m = \dfrac{-6 - 5}{4 - 2} = \dfrac{-11}{2}$ Find the slope of the line segment.
The slope of the lines perpendicular to the segment is $\dfrac{2}{11}$.

$m = \dfrac{2}{11}$, (2, 5)   Use the point-slope formula to find the equation of each of the two lines.

$y - 5 = \dfrac{2}{11}(x - 2)$
$11y - 55 = 2x - 4$
$2x - 11y = -51$

$m = \dfrac{2}{11}$, (4, -6)
$y - (-6) = \dfrac{2}{11}(x - 4)$
$11y + 66 = 2x - 8$
$2x - 11y = 74$

The equations of the lines are:
$2x - 11y = -51$
$2x - 11y = 74$

Solve by substitution:

41. $\begin{cases} 4x - 8y = -32 & (1) \\ x + 6y = -9 & (2) \end{cases}$

   $x + 6y = -9$ (2)   Solve equation (2) for x.
     $x = -6y - 9$

    $4x - 8y = -32$ (1)   Substitute $-6y - 9$ for x in equation (1).
$4(-6y - 9) - 8y = -32$
   $-32y - 36 = -32$
     $-32y = 4$
      $y = -\dfrac{1}{8}$

     $x + 6y = -9$ (2)   Substitute $-\dfrac{1}{8}$ for y in equation (2).
   $x + 6\left(-\dfrac{1}{8}\right) = -9$
     $8x - 6 = -72$ Multiply both sides by 8.
      $8x = -66$
       $x = -\dfrac{33}{4}$

The solution set is $\left\{\left(-\dfrac{33}{4}, -\dfrac{1}{8}\right)\right\}$.

EXERCISES 8.5 SOLVING SYSTEMS OF EQUATIONS INVOLVING
               QUADRATIC EQUATIONS I

A

Solve:

1. $\begin{cases} y = x^2 & (1) \\ y = 4x & (2) \end{cases}$

$y = 4x$ Equation (1) is solved for y.
$x^2 = 4x$ Substitute x^2 for y in
$x^2 - 4x = 0$ equation (2).
$x(x - 4) = 0$
$x = 0$ or $x = 4$

                                  Substitute both values of x
                                  in equation (1).
$y = 0^2 = 0$ $x = 0$
   (0, 0)   Ordered pair.
$y = 4^2 = 16$ $x = 4$
   (4, 16)   Ordered pair.

The solution set is $\{(0, 0), (4, 16)\}$.

5. $\begin{cases} x^2 + y^2 = 16 & (1) \\ x + y = 4 & (2) \end{cases}$

$x + y = 4$ Solve equation (2) for x.
$x = -y + 4$

$x^2 + y^2 = 16$ Substitute $-y + 4$ for x in
$(-y + 4)^2 + y^2 = 16$ equation (1).
$y^2 - 8y + 16 + y^2 = 16$
$2y^2 - 8y = 0$
$2y(y - 4) = 0$
$y = 0$ or $y = 4$

                                  Substitute both values of y
                                  in equation (2).
$x + 0 = 4$ $y = 0$
$x = 4$

   (0, 4)   Ordered pair.

The solution set is $\{(0, 4), (4, 0)\}$.

293

9. $\begin{cases} x^2 - y^2 = 1 & (1) \\ y = 2 & (2) \end{cases}$

$\quad x^2 - 2^2 = 1$ Equation (2) is solved for y,
$\quad\quad x^2 = 5$ so substitute 2 for y in
$\quad\quad\quad x = \pm\sqrt{5}$ equation (1).

The solution set is $\{(-\sqrt{5}, 2), (\sqrt{5}, 2)\}$.

B

13. $\begin{cases} x^2 - y^2 = 4 & (1) \\ x + y = 1 & (2) \end{cases}$

$\quad x + y = 1$ Solve equation (2) for x.
$\quad\quad x = -y + 1$

$\quad\quad\quad x^2 - y^2 = 4$ Substitute $-y + 1$ for x in
$\quad (-y + 1)^2 - y^2 = 4$ equation (1).
$\quad y^2 - 2y + 1 - y^2 = 4$
$\quad\quad\quad\quad -2y = 4$
$\quad\quad\quad\quad\quad y = -\dfrac{3}{2}$

$\quad x + y = 1$ Substitute $-\dfrac{3}{2}$ for y in
$\quad x - \dfrac{3}{2} = 1$ equation (2).
$\quad\quad x = \dfrac{5}{2}$

The solution set is $\left\{\left(\dfrac{5}{2}, -\dfrac{3}{2}\right)\right\}$.

17. $\begin{cases} y = x^2 - 4x + 2 & (1) \\ y - x = 2 & (2) \end{cases}$

$\quad y - x = 2$ Solve equation (2) for x.
$\quad y - 2 = x$

$\quad y = (y - 2)^2 - 4(y - 2) + 2$ Substitute $y - 2$ for x in
                                    equation (1).
$\quad y = y^2 - 4y + 4 - 4y + 8 + 2$
$\quad 0 = y^2 - 9y + 14$ Standard form.
$\quad 0 = (y - 7)(y - 2)$ Solve for y.
$\quad y = 7 \text{ or } y = 2$

$\quad y - 2 = x$ Substitute 7 for y in
$\quad 7 - 2 = x$ equation (2).
$\quad\quad 5 = x$

$\quad y - 2 = x$ Substitute 2 for y in
$\quad 2 - 2 = x$ equation (2).
$\quad\quad 0 = x$

The solution set is $\{(5, 7), (0, 2)\}$.

21. $\begin{cases} 25x^2 + y^2 = 25 & (1) \\ 5x + y = 5 & (2) \end{cases}$

$5x + y = 5$ Solve equation (2) for y.
$y = -5x + 5$

$25x^2 + (-5x + 5)^2 = 25$ Substitute $-5x + 5$ for y in
$25x^2 + 25x^2 - 50x + 25 = 25$ equation (1).
$50^2 - 50x = 0$
$50x(x - 1) = 0$
$x = 0$ or $x = 1$

$5(0) + y = 5$ Substitute both replacements
$y = 5$ for x in equation (2).

$5(1) + y = 5$
$y = 0$

The solution set is $\{(0, 5), (1, 0)\}$.

C

25. $\begin{cases} x^2 + y^2 = 6 & (1) \\ 2x - y = 6 & (2) \end{cases}$

$2x - y = 6$ Solve equation (1) for y.
$2x - 6 = y$

$x^2 + (2x - 6)^2 = 6$ Substitute $2x - 6$ for y in
$x^2 + 4x^2 - 24x + 36 = 6$ equation (1) and solve for x
$5x^2 - 24x + 30 = 0$ by using the quadratic
                                                formula.

$x = \dfrac{24 \pm \sqrt{576 - 4(5)(30)}}{10}$

$ = \dfrac{24 \pm \sqrt{-24}}{10}$

There are no real solutions. $\sqrt{-24}$ is not a real number.

29. $\begin{cases} y = -x^2 + 10x - 24 & (1) \\ x - y = 4 & (2) \end{cases}$

$\quad x - 4 = y$ Solve equation (2) for y.

$\quad\quad\quad x - 4 = -x^2 + 10x - 24$ Substitute $x - 4$ for y in equation (1).
$\quad x^2 - 9x + 20 = 0$ Solve for x.
$\quad (x - 5)(x - 4) = 0$
$\quad x = 5 \text{ or } x = 4$

$\quad 5 - y = 4$ Substitute both replacements
$\quad\quad 1 = y$ for x in equation (2).

$\quad 4 - y = 4$
$\quad\quad 0 = y$

The solution set is $\{(5, 1), (4, 0)\}$.

D

33. A rectangle has an area of 3200 cm² and a perimeter of 240 cm. What are its dimensions?

 Select variables:
 Let x represent the length and let y represent the width.

 Formulas:
 $\quad xy = A$
 $2x + 2y = P$

 Solve:
 $\quad xy = 3200$ (1)
 $2x + 2y = 240$ (2)

 $\quad\quad 2x = 240 - 2y$ Solve equation (2) for x.
 $\quad\quad\quad x = 120 - y$

 $(120 - y)(y) = 3200$ Substitute $120 - y$ for x
 $\quad 120y - y^2 = 3200$ in equation (1).
 $\quad\quad\quad\quad 0 = y^2 - 120y + 3200$
 $\quad\quad\quad\quad 0 = (y - 40)(y - 80)$
 $\quad y = 40 \text{ or } y = 80$

 $x(40) = 3200$ Substitute $y = 40$ in
 $\quad\quad x = 80$ equation (1).

 $x(80) = 3200$ Substitute $y = 80$ in
 $\quad\quad x = 40$ equation (1).

 The solution set is $\{(40, 80), (80, 40)\}$.

 Answer:
 The rectangle is 40 cm × 80 cm.

E Maintain Your Skills

Evaluate:

37. $\begin{vmatrix} 6 & 4 \\ 6 & -3 \end{vmatrix}$

$\begin{vmatrix} 6 & 4 \\ 6 & -3 \end{vmatrix} = 6(-3) - 6(4)$ Find the products indicated
$= -18 - 24$ by the arrows. Subtract
$= -42$ these products.

41. $\begin{vmatrix} 11 & -3 & 4 \\ 1 & -2 & 0 \\ 6 & 1 & 3 \end{vmatrix}$

$\begin{vmatrix} 11 & -3 & 4 \\ 1 & -2 & 0 \\ 6 & 1 & 3 \end{vmatrix} = \begin{vmatrix} -2 & 0 \\ 1 & 3 \end{vmatrix} - 1 \begin{vmatrix} -3 & 4 \\ 1 & 3 \end{vmatrix} + 6 \begin{vmatrix} -3 & 4 \\ -2 & 0 \end{vmatrix}$

$= 11(-6 - 0) - 1(-9 - 4) + 6[0, -(-8)]$
$= -66 + 13 + 48$
$= -5$

45. Sonia takes 6 hours to row 12 miles upstream and 3 hours to row back down to her starting Point. Find Sonia's average rowing rate in still water and the rate of the current of the stream.

<u>Select variables:</u>
Let x represent Sonia's average rowing rate in still water. Let y represent the rate of the current in the stream.

<u>Simpler word form:</u>

$\begin{bmatrix} \text{Sonia's} \\ \text{average} \\ \text{rowing rate} \end{bmatrix} - \begin{bmatrix} \text{Rate of} \\ \text{current} \\ \text{in stream} \end{bmatrix} = \begin{bmatrix} \text{Rate} \\ \text{going} \\ \text{upstream} \end{bmatrix}$

$\begin{bmatrix} \text{Sonia's} \\ \text{average} \\ \text{rowing rate} \end{bmatrix} - \begin{bmatrix} \text{Rate of} \\ \text{current} \\ \text{in stream} \end{bmatrix} = \begin{bmatrix} \text{Rate} \\ \text{going} \\ \text{downstream} \end{bmatrix}$

$R = \frac{D}{T}$ Find the rate going upstream.
$= \frac{12}{6} = 2$

$R = \frac{D}{T}$ Find the rate going downstream.
$= \frac{12}{3} = 4$

<u>Translate to algebra:</u>
$x - y = 2$
$x + y = 4$

Solve:

$$D = \begin{vmatrix} 1 & -1 \\ 1 & 1 \end{vmatrix} \qquad D_x = \begin{vmatrix} 2 & -1 \\ 4 & 1 \end{vmatrix}$$

Write determinants D, D_x, and D_y.

$$D_y = \begin{vmatrix} 1 & 2 \\ 1 & 4 \end{vmatrix}$$

Expand D, D_x, and D_y.

$D = 1 - (-1) = 2$
$D_x = 2 - (-4) = 6$
$D_y = 4 - 2 = 2$

$x = \dfrac{D_x}{D} = \dfrac{6}{2} = 3$

$y = \dfrac{D_y}{D} = \dfrac{2}{2} = 1$

Answer:
Sonia's average rowing rate is 3 mph, and the rate of the current is 1 mph.

EXERCISES 8.6 SOLVING SYSTEMS OF EQUATIONS INVOLVING QUADRATIC EQUATIONS II

A

Solve:

1. $\begin{cases} x^2 + y^2 = 12 & (1) \\ x^2 - y^2 = 4 & (2) \end{cases}$

$\begin{aligned} x^2 + y^2 &= 12 \\ \underline{x^2 - y^2 = 4} \\ 2x^2 &= 16 \\ x^2 &= 8 \\ x &= \pm\sqrt{8} = \pm 2\sqrt{2} \end{aligned}$

Add equations (1) and (2).

$(\pm 2\sqrt{2})^2 + y^2 = 12$
$8 + y^2 = 12$
$y^2 = 4$
$y = \pm 2$

Substitute $x = \pm 2\sqrt{2}$ in equation (1).

The solution set is $\{(2\sqrt{2}, 2), (2\sqrt{2}, -2), (-2\sqrt{2}, 2), (-2\sqrt{2}, -2)\}$.

5. $\begin{cases} x^2 + y^2 = 4 & (1) \\ 4x^2 + y^2 = 4 & (2) \end{cases}$

$\begin{aligned} -x^2 - y^2 &= -4 \\ \underline{4x^2 + y^2} &= \underline{4} \\ 3x^2 &= 0 \\ x &= 0 \end{aligned}$ 	Multiply equation (1) by -1, and form a combination by adding the result to equation (2).

$\begin{aligned} 0^2 + y^2 &= 4 \\ y^2 &= 4 \\ y &= \pm 2 \end{aligned}$ 	Substitute $x = 0$ in equation (1).

The solution set is $\{(0, 2), (0, -2)\}$.

B

9. $\begin{cases} 2x^2 + y^2 = 6 & (1) \\ 4x^2 + 3y^2 = 16 & (2) \end{cases}$

$\begin{aligned} -4x^2 - 2y^2 &= -12 \\ \underline{4x^2 + 3y^2} &= \underline{16} \\ y^2 &= 4 \\ y &= \pm 2 \end{aligned}$ 	Form a combination by multiplying equation (1) by -2 and adding to equation (2). The terms containing x^2 will disappear.

$\begin{aligned} 2x^2 + (\pm 2)^2 &= 6 \\ 2x^2 + 4 &= 6 \\ 2x^2 &= 2 \\ x^2 &= 1 \\ x &= \pm 1 \end{aligned}$ 	Substitute $y = \pm 2$ in equation (1).

The solution set is $\{(1, 2), (1, -2), (-1, 2), (-1, -2)\}$.

13. $\begin{cases} x^2 + y^2 = 9 & (1) \\ 16x^2 + y^2 = 16 & (2) \end{cases}$

$\begin{aligned} -x^2 - y^2 &= -9 \\ \underline{16x^2 + y^2} &= \underline{16} \\ 15x^2 &= 7 \\ x^2 &= \frac{7}{15} \\ x &= \pm\sqrt{\frac{7}{15}} \end{aligned}$

Form a combination by multiplying equation (1) by -1 and adding it to equation (2).

$\pm\sqrt{\frac{7}{15}} = \pm\frac{\sqrt{7}}{\sqrt{15}} \cdot \frac{\sqrt{15}}{\sqrt{15}}$

Rationalize the denominator to simplify.

$\left[\pm\frac{\sqrt{105}}{15}\right]^2 + y^2 = 9$

Substitute $x = \frac{\sqrt{105}}{15}$ and

$\frac{105}{225} + y^2 = 9$

$x = -\frac{\sqrt{105}}{15}$ in equation (1).

$\frac{7}{15} + y^2 = 9$

Reduce the fraction.

$y^2 = \frac{135 - 7}{15}$

$y^2 = \frac{128}{15}$

$y = \pm\sqrt{\frac{128}{15}}$

$\pm\sqrt{\frac{128}{15}} = \pm\frac{\sqrt{128}}{\sqrt{15}} \cdot \frac{\sqrt{15}}{\sqrt{15}} = \pm\frac{\sqrt{1920}}{15}$

Rationalize the denominator to simplify.

$= \pm\frac{8\sqrt{30}}{15}$

So the solution set is

$\left\{\left(\frac{\sqrt{105}}{15}, \frac{8\sqrt{30}}{15}\right), \left(-\frac{\sqrt{105}}{15}, \frac{8\sqrt{30}}{15}\right), \left(\frac{\sqrt{105}}{15}, -\frac{8\sqrt{30}}{15}\right), \left(-\frac{\sqrt{105}}{15}, -\frac{8\sqrt{30}}{15}\right)\right\}.$

17. $\begin{cases} 3x^2 + 3y^2 = 36 & (1) \\ 2x^2 + 2y^2 = 4 & (2) \end{cases}$

$\begin{aligned} -x^2 - y^2 &= -12 \\ \underline{x^2 + y^2} &= \underline{2} \\ 0 &= -10 \end{aligned}$

Form a combination. Multiply equation (1) by $-\frac{1}{3}$ and equation (2) by $\frac{1}{2}$. Add.

Contradiction.

There are no real solutions.

300

21. $x^2 - 4y^2 = 19$ (1)
$x^2 - \frac{1}{2}y^2 = 26$ (2)

$\begin{aligned}-2x^2 + 8y^2 &= -38\\ 2x^2 - y^2 &= \\ \hline 7y^2 &= 14\\ y^2 &= 2\\ y &= \pm\sqrt{2}\end{aligned}$

Form a combination. Multiply equation (1) by -2, and equation (2) by 2.

$\begin{aligned}x^2 - 4(\pm\sqrt{2})^2 &= 19\\ x^2 - 8 &= 19\\ x^2 &= 27\\ x &= \pm\sqrt{27} = \pm 3\sqrt{3}\end{aligned}$

Substitute $\pm\sqrt{2}$ for y in equation (1).

The solution set is
$\{(3\sqrt{3}, \sqrt{2}), (3\sqrt{3}, -\sqrt{2}), (-3\sqrt{3}, \sqrt{2}), (-3\sqrt{3}, -\sqrt{2})\}$.

25. $\begin{cases} x^2 + y^2 = 10 & (1)\\ x^2 = 2y - 5 & (2)\end{cases}$

$\begin{aligned}2y - 5 + y^2 &= 10\\ y^2 + 2y - 15 &= 0\\ (y + 5)(y - 3) &= 0\\ y = -5 \text{ or } y &= 3\end{aligned}$

Substitute $2y - 5$ for x^2 in equation (1).

Zero-product property.

$\begin{aligned}x^2 &= 2(-5) - 5\\ x^2 &= -15\\ x &= \pm\sqrt{-15}\end{aligned}$

Substitute $y = -5$ and $y = 3$ in equation (2).

No real solution because $\sqrt{-15}$ is a complex number.

$\begin{aligned}x^2 &= 2(3) - 5\\ x^2 &= 1\\ x &= \pm 1\end{aligned}$

The solution set is $\{(1, 3), (-1, 3)\}$.

D

29. The sum of the squares of two numbers is 113. When 5 times the square of one is added to the square of the other, the sum is 309. What are the numbers?

$\begin{cases} x^2 + y^2 = 113 & (1) \\ 5x^2 + y^2 = 309 & (2) \end{cases}$ Form a system of equations with the given data.

$\begin{aligned} -x^2 - y^2 &= -113 \\ \underline{5x^2 + y^2} &= \underline{309} \\ 4x^2 &= 196 \\ x^2 &= 49 \\ x &= \pm 7 \end{aligned}$ Multiply equation (1) by -1, and add to equation (2).

$\begin{aligned} (\pm 7)^2 + y^2 &= 113 \\ y^2 &= 113 - 49 \\ y^2 &= 64 \\ y &= \pm 8 \end{aligned}$ Substitute ± 7 for x in equation (1) and solve for y.

The solution set is $\{(7, 8), (-7, 8), (7, -8), (-7, -8)\}$.

The numbers are 7 and 8, or -7 and -8, or 7 and -8, or -7 and 8.

E Maintain Your Skills

Simplify. Reduce, if possible:

33. $\dfrac{x^2 + x - 30}{2x^2 + 13x - 7} \cdot \dfrac{3x^2 + 22x + 7}{5x^2 - 31x + 30} \cdot \dfrac{10x^2 + 7x - 6}{x^2 + 7x + 6}$

$= \dfrac{(\cancel{x+6})(\cancel{x-5})}{(2x-1)(\cancel{x+7})} \cdot \dfrac{(3x+1)(\cancel{x+7})}{(5x-6)(\cancel{x-5})} \cdot \dfrac{(5x+6)(2x\cancel{-1})}{(x+1)(\cancel{x+6})}$

   Factor and divide out common factors.

$= \dfrac{(3x+1)(5x+6)}{(5x-6)(x+1)}$

Solve by any method:

37. $\begin{cases} x + y + z = 8 & (1) \\ x - y + 2z = 12 & (2) \\ 2x + 4y - z = 6 & (3) \end{cases}$

$\begin{array}{r} x + y + z = 8 \\ \underline{x - y + 2z = 12} \\ 2x + 3z = 20 \quad (4) \end{array}$ Add equations (1) and (2).

$\begin{array}{r} -4x - 4y - 4z = -32 \\ \underline{2x + 4y - z = 6} \\ -2x - 5z = -26 \quad (5) \end{array}$ Multiply equation (1) by −4 and add to equation (3).

$\begin{array}{r} 2x + 3z = 20 \\ \underline{-2x - 5z = -26} \\ -2z = -6 \\ z = 3 \end{array}$ Add equations (4) and (5).

$\begin{array}{r} 2x + 3(3) = 20 \\ 2x = 11 \\ x = \dfrac{11}{2} \end{array}$ Replace z in equation (4) with 3.

$\dfrac{11}{2} + y + 3 = 8$ Substitute $\dfrac{11}{2}$ for x and 3 for z in equation (1). Solve for y.

$11 + 2y + 6 = 16$
$2y = -1$
$y = -\dfrac{1}{2}$

The solution set is $\left\{ \left[\dfrac{11}{2}, -\dfrac{1}{2}, 3 \right] \right\}$.

CHAPTER 9

EXPONENTIAL AND LOGARITHMIC FUNCTIONS

EXERCISES 9.1   EXPONENTIAL FUNCTIONS

A

Complete the following table of values:

1. $f(x) = 3^x$

x	-3	-2	-1	0	1	2	3
f(x)	$\frac{1}{27}$	$\frac{1}{9}$	$\frac{1}{3}$	1	3	9	27

$f(-3) = 3^{-3} = \dfrac{1}{3^3} = \dfrac{1}{27}$

$f(-2) = 3^{-2} = \dfrac{1}{3^2} = \dfrac{1}{9}$

$f(-1) = 3^{-1} = \dfrac{1}{3^1} = \dfrac{1}{3}$

$f(0) \ = 3^0 = 1$

$f(1) \ = 3^1 = 3$

$f(2) \ = 3^2 = 9$

$f(3) \ = 3^3 = 27$

5. $f(x) = \left(\dfrac{2}{3}\right)^x$

x	-3	-2	-1	0	1	2	3
f(x)	$\dfrac{27}{8}$	$\dfrac{9}{4}$	$\dfrac{3}{2}$	1	$\dfrac{2}{3}$	$\dfrac{4}{9}$	$\dfrac{8}{27}$

$f(-3) = \left(\dfrac{2}{3}\right)^{-3} = \left(\dfrac{3}{2}\right)^3 = \dfrac{27}{8}$

$f(-2) = \left(\dfrac{2}{3}\right)^{-2} = \left(\dfrac{3}{2}\right)^2 = \dfrac{9}{4}$

$f(-1) = \left(\dfrac{2}{3}\right)^{-1} = \left(\dfrac{3}{2}\right)^1 = \dfrac{3}{2}$

$f(0) = \left(\dfrac{2}{3}\right)^0 = 1$

$f(1) = \left(\dfrac{2}{3}\right)^1 = \dfrac{2}{3}$

$f(2) = \left(\dfrac{2}{3}\right)^2 = \dfrac{4}{9}$

$f(3) = \left(\dfrac{2}{3}\right)^3 = \dfrac{8}{27}$

9. $f(x) = 10^x$

x	-3	-2	-1	0	1	2
f(x)	$\dfrac{1}{1000}$	$\dfrac{1}{100}$	$\dfrac{1}{10}$	1	10	100

Make a table of values.

$f(-3) = 10^{-3} = \dfrac{1}{10^3} = \dfrac{1}{1000}$

$f(-2) = 10^{-2} = \dfrac{1}{10^2} = \dfrac{1}{100}$

$f(-1) = 10^{-1} = \dfrac{1}{10^1} = \dfrac{1}{10}$

$f(0) = 10^0 = 1$

$f(1) = 10^1 = 10$

$f(2) = 10^2 = 100$

B

Use the table of Exercise 1 to graph the function defined by the following equation. Write the domain and range for the function:

13. $f(x) = 3^x$

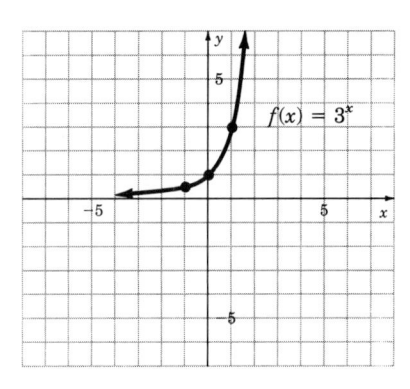

Sketch the graph by plotting the points from the table of values of Exercise 1.

Domain: $\{x \mid x \in \mathbb{R}\}$ or $\mathbb{R}$

Range: $\{y \mid y > 0\}$ or $\mathbb{R}^+$

17. $f(x) = \left(\dfrac{3}{4}\right)^x$

x	-3	-2	-1	0	1	2	3
f(x)	$\dfrac{64}{27}$	$\dfrac{16}{9}$	$\dfrac{4}{3}$	1	$\dfrac{3}{4}$	$\dfrac{9}{16}$	$\dfrac{27}{64}$

Make a table of values.

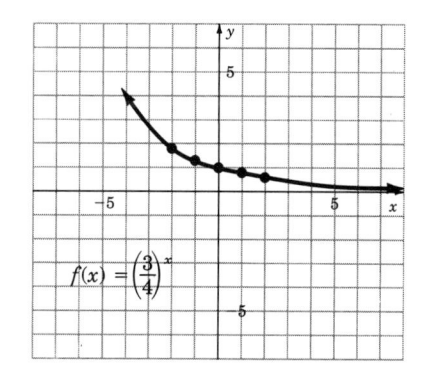

Sketch the graph by connecting the points with a smooth curve.

Domain: $\{x \mid x \in \mathbb{R}\}$

Range: $\{y \mid y > 0\}$

C

Graph the function defined by each of these equations. Write the domain and range of each:

21. $f(x) = \left(\dfrac{3}{2}\right)^x$

x	-3	-2	-1	0	1	2	3
f(x)	$\dfrac{8}{27}$	$\dfrac{4}{9}$	$\dfrac{2}{3}$	1	$\dfrac{3}{2}$	$\dfrac{9}{4}$	$\dfrac{27}{8}$

Make a table of values.

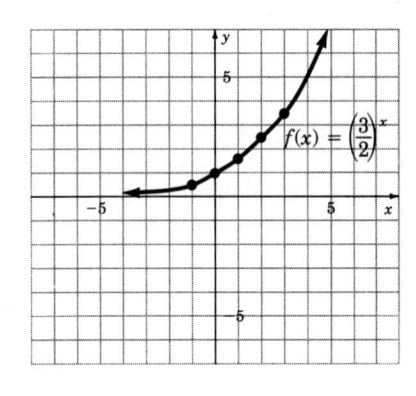

Sketch the graph.

Domain: $\{x \mid x \in \mathbb{R}\}$
Range: $\{y \mid y > 0\}$

25. $f(x) = (1.2)^x$

x	-2	-1	0	1	2	3	4
f(x)	0.69	0.83	1	1.2	1.44	1.73	2.07

Make a table of values.

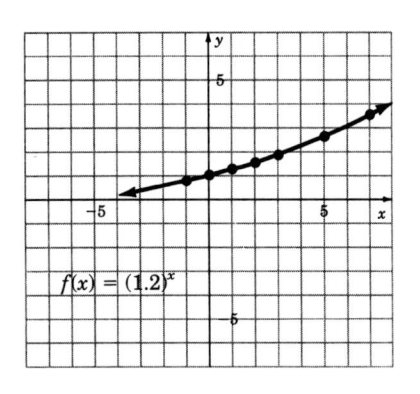

Sketch the graph.

Domain: $\{x \mid x \in \mathbb{R}\}$
Range: $\{y \mid y > 0\}$

D

29. Based on the growth of the population for the past 20 years, the formula for the future world population is (approximately) $P = 4e^{0.2t} \times 10^9$, where t is the number of years and $e \approx 2.718$. What is the population now (t = 0)? What will the population be (approximately) 10 years from now?

$P = 4e^{0.2t} \times 10^9$	Formula.
$P = 4e^{(0.2)0} \times 10^9$	Substitute t = 0.
$\quad = 4e^0 \times 10^9$	
$\quad = 4 \times 10^9$	$e^0 = 1$.
$\quad = 4,000,000,000$	
$P = 4e^{(0.2)(10)} \times 10^9$	Substitute t = 10
$\quad = 4e^2 \times 10^9$	
$\quad \approx 4(7.4) \times 10^9$	$e^2 \approx 7.4$
$\quad \approx 29.6 \times 10^9$	
$\quad \approx 29,600,000,000$	

The population now is approximately 4,000,000,000. The population 10 years from now will be approximately 29,600,000,000.

33. The formula for interest compounded quarterly is $A = \left[1 + \frac{r}{4}\right]^n$, where A is the total amount on deposit after n quarter-year periods if P dollars are invested at 4 percent per year.

Ms. Skinflint invests $1,000 at 8% interest compounded quarterly, which is to be given to her first great-grandchild upon its twenty-first birthday. If the great-granddaughter is 21 years of age 40 years after the money is invested (160 periods), how much does she receive? (Round to the nearest dollar.)

$A = \left[1 + \frac{r}{4}\right]^n$	Formula.
$\quad = 1000\left[1 + \frac{0.08}{4}\right]^{160}$	Substitute 1000 for P, 0.08 (8%) for r, and 160 for n.
$\quad = 1000(1 + 0.02)^{160}$	
$\quad = 1000(1.02)^{160}$	

ENTER

| 1000 | × | 1.02 | y^x | 160 | = |

Use a calculator.

DISPLAY

| 23769.906 |

≈ 23770. Round to the nearest whole number.

She received approximately $23,770 from Ms. Skinflint.

E Maintain Your Skills

Multiply and simplify:

37. $(\sqrt{3} + \sqrt{6})(\sqrt{3} - \sqrt{10}) = \sqrt{9} - \sqrt{30} + \sqrt{18} - \sqrt{60}$
$= 3 - \sqrt{30} + 3\sqrt{2} - 2\sqrt{15}$

Rationalize the denominator:

41. $\dfrac{\sqrt{5} + 2}{3 - \sqrt{5}} = \dfrac{(\sqrt{5} + 2)(3 + \sqrt{5})}{(3 - \sqrt{5})(3 + \sqrt{5})}$

$= \dfrac{3\sqrt{5} + \sqrt{25} + 6 + 2\sqrt{5}}{9 - \sqrt{25}}$

$= \dfrac{11 + 5\sqrt{5}}{4}$

Solve using determinants:

45. $\begin{aligned} 2x + y &= 8 &(1) \\ x + y + z &= 7 &(2) \\ 4y - 5z &= 11 &(3) \end{aligned}$

Write the determinants D, D_x, D_y, and D_z.

$D = \begin{vmatrix} 2 & 1 & 0 \\ 1 & 1 & 1 \\ 0 & 4 & -5 \end{vmatrix} = 2(-9) - 1(-5) + 0(1) = -13$ Expand about the first column.

$D_x = \begin{vmatrix} 8 & 1 & 0 \\ 7 & 1 & 1 \\ 11 & 4 & -5 \end{vmatrix} = 8(-9) - 1(-46) + 0(17) = -26$ Expand about the first row.

$$D_y = \begin{vmatrix} 2 & 8 & 0 \\ 1 & 7 & 1 \\ 0 & 11 & -5 \end{vmatrix} = 2(-46) - 1(-40) + 0(8) = -52 \quad \text{Expand about the first column.}$$

$$D_z = \begin{vmatrix} 2 & 1 & 8 \\ 1 & 1 & 7 \\ 0 & 4 & 11 \end{vmatrix} = 2(-17) - 1(-21) + 0(-1) = -13 \quad \text{Expand about the first column.}$$

$x = \dfrac{D_x}{D} = \dfrac{-26}{-13} = 2$ Use Cramer's rule to solve.

$y = \dfrac{D_y}{D} = \dfrac{-52}{-13} = 4$

$z = \dfrac{D_z}{D} = \dfrac{-13}{-13} = 1$

The solution set is {(2, 4, 1)}.

EXERCISES 9.2   LOGARITHMIC FUNCTIONS

A

Write each logarithmic equation in exponential form:

1. $\log_3 27 = 3$
   $3^3 = 27$

   3 is the base.
   3 is the exponent or logarithm.
   27 is the power.

5. $\log_x x^2 = 2$
   $x^2 = x^2$

   x is the ase.
   2 is the exponent.
   x^2 is the power.

9. $\log_{1/2} \dfrac{1}{8} = 3$
   $\left(\dfrac{1}{2}\right)^3 = \dfrac{1}{8}$

   $\dfrac{1}{2}$ is the base.
   3 is the exponent.
   $\dfrac{1}{8}$ is the power.

Write each exponential equation in logarithmic form:

13. $2^6 = 64$
     $\log_2 64 = 6$

   2 is the base.
   6 is the logarithm or exponent.
   64 is the ower.

311

17. $8^{1/2} = \sqrt{8}$

       $\log_8 \sqrt{8} = \frac{1}{2}$ 8 is the base.

                                       $\frac{1}{2}$ is the logarithm.

                                       $\sqrt{8}$ is the power.

B

Find w:

21. $w = \log_6 36$

       $6^w = 36$ Change from log form to exponential form.

       $6^w = 6^2$

       $w = 2$ If $a^x = a^y$, then $x = y$.

25. $w = \log_{10} \frac{1}{10}$

       $10^w = \frac{1}{10}$ Change from log form to exponential form.

       $10^w = 10^{-1}$

       $w = -1$ If $a^x = a^y$, then $x = y$.

29. $w = \log_5 \frac{1}{625}$

       $5^w = \frac{1}{625}$ Change from log form to exponential form.

       $5^w = 5^{-4}$

       $w = -4$ If $a^x = a^y$, then $x = y$.

33. $w = \log_{1/4} 16$

       $\left(\frac{1}{4}\right)^w = 16$ Change from log form to exponential form.

       $\left(\frac{1}{4}\right)^w = \left(\frac{1}{4}\right)^{-2}$

       $w = -2$

Complete the table of values:

37. $y = \log_{10} x$

x	$\frac{1}{100}$	$\frac{1}{10}$	1	100	1000
y					

Change $y = \log_{10} x$ to exponential form:
$x = 10^y$ and fill in the table of values.

$x = 10^y$

$x = \frac{1}{100}$ $\qquad$ $x = \frac{1}{10}$ $\qquad$ $x = 1$
$\frac{1}{100} = 10^y$ $\qquad$ $\frac{1}{10} = 10^y$ $\qquad$ $1 = 10^y$
$10^{-2} = 10^y$ $\qquad$ $10^{-1} = 10^y$ $\qquad$ $10^0 = 10^y$
$y = -2$ $\qquad$ $y = -1$ $\qquad$ $y = 0$

$x = 100$ $\qquad$ $x = 1000$
$100 = 10^y$ $\qquad$ $1000 = 10^y$
$10^2 = 10^y$ $\qquad$ $10^3 = 10^y$
$y = 2$ $\qquad$ $y = 3$

x	$\frac{1}{100}$	$\frac{1}{10}$	1	100	1000
y	-2	-1	0	2	3

C

Find z:

41. $z = \log_4 8$
$4^z = 8$
$(2^2)^z = 2^3$ $\qquad$ Change to exponential form. Rewrite both sides using base 2.
$2^{2z} = 2^3$
$2z = 3$
$z = \frac{3}{2}$ $\qquad$ If $a^x = a^y$, then $x = y$.

313

45. $z = \log_{32} 16$

       $32^z = 16^{32}$ Change to exponential form.

       $2^{5z} = 2^4$ Rewrite both sides using base 2.

        $5z = 4$

          $z = \dfrac{4}{5}$

49. $z = \log_{3/2}\left(\dfrac{8}{27}\right)$

     $\left(\dfrac{3}{2}\right)^z = \dfrac{8}{27}$ Change to exponential form.

     $\left(\dfrac{3}{2}\right)^z = \left(\dfrac{3}{2}\right)^{-3}$ Rewrite both sides using base $\dfrac{3}{2}$.

         $z = -3$

Use the tables in Exercises 35 and 39 to sketch the following; state the domain and range:

53. $y = \log_4 x$

x	$\dfrac{1}{16}$	$\dfrac{1}{4}$	1	4	64
y	-2	-1	0	1	3

Table of values.

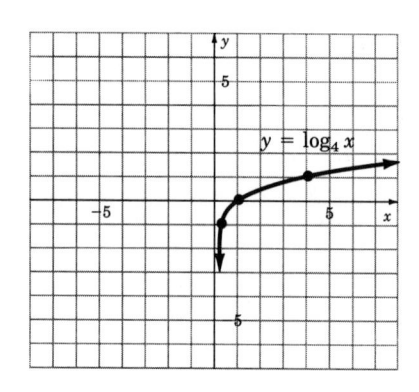

Sketch the graph by plotting the points and connecting them with a smooth curve.

Domain: $\{x \mid x > 0\}$ or $\mathbb{R}^+$
Range: $\{y \mid y \in \mathbb{R}\}$ or $\mathbb{R}$

57. $y = \log_{1/2} x$

x	4	2	1	$\frac{1}{2}$	$\frac{1}{4}$
y	-2	-1	0	1	2

Table of values.

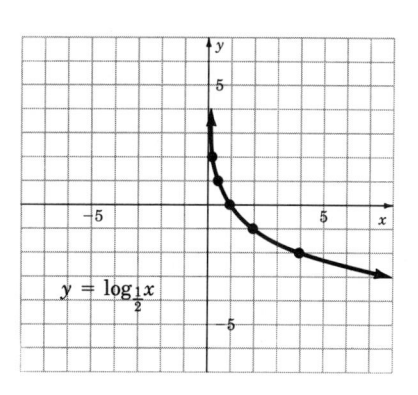

Sketch the graph.

Domain: $\{x | x > 0\}$ or $\mathbb{R}^+$
Range: $\{y | y \in \mathbb{R}\}$ or $\mathbb{R}$

61. Find the pH of a solution with an H^+ value of 1×10^5.

$$pH = \log_{10}\left[\frac{1}{H^+}\right]$$ Formula.

$$pH = \log_{10}\left[\frac{1}{1 \times 10^5}\right]$$ Substitute.

$$pH = \log_{10} 10^{-5}$$ Solve. $\frac{1}{1 \times 10^5} = \frac{1}{10^5} = 10^{-5}$.

$$10^{pH} = 10^{-5}$$ Write in exponential form.

$$pH = -5$$

The pH of the solution is -5.

315

E Maintain Your Skills

Solve:

65. $0 = -\frac{1}{2}(x - 6)^2 + 2$

$-2 = -\frac{1}{2}(x - 6)^2$ Subtract 2 from both sides.

$4 = (x - 6)^2$ Multiply both sides by -2.

$\pm 2 = x - 6$ Extract the square root on both sides.

$6 \pm 2 = x$

$x = 6 + 2$ or $6 - 2$

$x = 8$ or 4

The solution set is $\{4, 8\}$.

69. What is the maximum value of the function $f(x) = -\frac{1}{2}(x - 6)^2 + 2$?

$a = -\frac{1}{2}$, $h = 6$, $k = 2$ The coordinates of the maximum point are (h, k).

$(h, k) \rightarrow (6, 2)$

So $f(6) = 2$.   The value of the function at $x = 6$ is 2.

73. What is the minimum value of the function $f(x) = 2x^2 + 8x - 3$?

$a = 2$, $b = 8$, and $c = -3$ The x-coordinate of the vertex is $-\frac{b}{2a}$.

$-\frac{b}{2a} = -\frac{8}{4} = -2$

$f(-2) = 2(-2)^2 + 8(-2) - 3$ Substitute -2 for x in $f(x)$.
$= 2(4) - 16 - 3$
$= 8 - 16 - 3$
$= -11$

So $f(-2) = -11$.   The value of the function at $x = -2$ is -11.

EXERCISES 9.3   PROPERTIES OF LOGARITHMS

A

Write the following as logarithms of a single number:

1. $\log_b 6 + \log_b 8 = \log_b 48$ Property Five of logarithms.

5. $2\log_6 8 = \log_6 8^2 = \log_6 64$ Property Seven of logarithms.

316

9. $-2\log_4 3 = \log_4 3^{-2}$ Property Seven.
           $= \log_4 \left[\dfrac{1}{9}\right]$

Write the following as a combination of logarithms of a single number:

13. $\log_b (132)^4 = 4\log_b 132$ Property Seven of logarithms.

B

Write the following as logarithms of single numbers:

17. $3\log_b 4 + \log_b 3$
       $= \log_b 4^3 + \log_b 3$ Property Seven of logarithms.
       $= \log_b (4^3 \cdot 3)$ Property Five of logarithms.
       $= \log_b 192$

21. $2\log_b 5 + \log_b 2 - \log_b 10$
       $= \log_b 5^2 + \log_b 2 - \log_b 10$ Property Seven of logarithms.
       $= \log_b (5^2 \cdot 2) - \log_b 10$ Property Five.
       $= \log_b \left[\dfrac{5^2 \cdot 2}{10}\right] = \log_b \left[\dfrac{50}{10}\right] = \log_b 5$ Property Six. Simplify.

Write the following as combinations of logarithms of single variables:

25. $\log_b \left[\dfrac{z}{x}\right] = \log_b z - \log_b x$ Property Six.

29. $\log_5 \dfrac{x^2}{y^2} = \log_5 x^2 - \log_5 y^2$ Property Six.

         $= 2\log_5 x - 2\log_5 y$ Property Seven.

C

Write the following as logarithms of single numbers:

33. $2\log_b 4 + 3\log_b 2 - 3\log_b 4$
       $= \log_b 4^2 + \log_b 2^3 - \log_b 4^3$ Property Seven.
       $= \log_b (4^2 \cdot 2^3) - \log_b 4^3$ Property Five.
       $= \log_b \left[\dfrac{4^2 \cdot 2^3}{4^3}\right] = \log_b \left[\dfrac{128}{64}\right] = \log_b 2$ Property Six. Simplify.

37. $\frac{1}{3}\log_b 8 + \frac{1}{2}\log_b 49$

  $= \log_b 8^{1/3} + \log_b 49^{1/2}$ Property Seven.

  $= \log_b (8^{1/3} \cdot 49^{1/2})$ Property Five.

  $= \log_b (2 \cdot 7) = \log_b 14$ Simplify.

Write the following as combinations of logarithms of single variables:

41. $\log_c z\sqrt{xy} = \log_c z(xy)^{1/2}$

     $= \log_c zx^{1/2}y^{1/2}$

     $= \log_c z + \log_c x^{1/2} + \log_c y^{1/2}$

     $= \log_c z + \frac{1}{2}\log_c x + \frac{1}{2}\log_c y$

45. $\log_x \sqrt[3]{(a+b)^2} = \log_x (a+b)^{2/3}$

      $= \frac{2}{3}\log_x (a+b)$

D

49. What is the magnitude on the Richter scale of an earthquake that is $10^{2.5}$ times stronger than the zero quake? (Use the formula $M(i) = \log_{10}\frac{i}{i_0}$ where is the amplitude of the ground motion of the quake and i_0 is the amplitude of the ground motion of a "zero" quake.

Formula:
$M(i) = \log_{10}\frac{i}{i_0}$

Substitute:
$M(10^{2.5}) = \log_{10}\frac{10^{2.5}i_0}{i_0}$ $i = 10^{2.5}$

    $= \log_{10}10^{2.5}$ Reduce.

    $= 2.5\log_{10}10$

    $= 2.5$ $\log_{10}10 = 1$

Answer:
The magnitude of the earthquake is 2.5.

53. Find the pH of a solution of hydrochloric acid (a strong acid) if its hydrogen ion concentration (H^+) is 5×10^{-2}. (Use $\log_{10} 0.2 \approx -0.6990$, and round to the nearest tenth.) The formula for pH is $pH = \log_{10} \frac{1}{H^+}$.

Formula:
$$pH = \log_{10} \frac{1}{H^+}$$

Substitute:
$$pH = \log_{10} \frac{1}{5 \times 10^{-2}} \qquad H^+ \approx 5 \times 10^{-2}$$

Solve:
$$pH = \log_{10} \left[\frac{1}{5} \cdot \frac{1}{10^{-2}} \right]$$

$$= \log_{10} \frac{1}{5} + \log_{10} \frac{1}{10^{-2}}$$

$$= \log_{10} 0.2 + \log_{10} 10^2 \qquad \text{Substitute } \frac{1}{5} = -0.2 \text{ and } \frac{1}{10^{-2}} = 10^2.$$

$$= \log_{10} 0.2 + 2 \log_{10} 10$$

$$= -0.6990 + 2 \qquad \text{Substitute } \log_{10} 0.2 = -0.6990.$$
$$= 1.301 \qquad \text{(given) and } \log_{10} 10 = 1.$$
$$\approx 1.3 \text{ (to the nearest tenth)}$$

Answer:
The pH of the solution of hydrochloric acid is approximately 1.3.

E Maintain Your Skills

57. Find the radius of the circle:

$$x^2 + 6x + y^2 - 1 = 0$$

Write the equation in standard form.
Complete the square in both x and y.

$$x^2 + 6x + y^2 = 1$$

$$x^2 + 6x + 9 + (y + 0)^2 = 1 + 9$$

$$(x + 3)^2 + (y + 0)^2 = 10 \qquad r^2 = 10, \; r = \sqrt{10}.$$

The radius is $\sqrt{10}$.

61. Identify the conic section whose equation is $9x^2 - 25y^2 = 144$ and graph.

$9x^2 - 25y^2 = 144$

$\dfrac{9x^2}{144} - \dfrac{25y^2}{144} = 1$ Write in standard form. Divide both sides by 144.

$\dfrac{x^2}{16} - \dfrac{25y^2}{144} = 1$

$\dfrac{x^2}{16} - \dfrac{y^2}{\frac{144}{25}} = 1$ This is the standard form of a hyperbola, with intercepts at (4, 0) and (-4, 0).

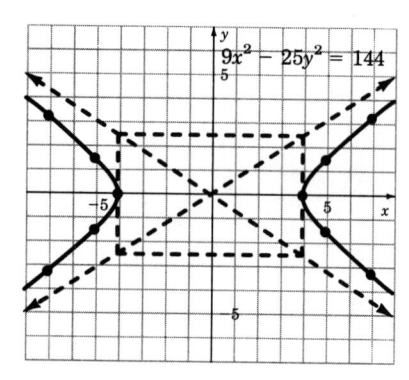

Table of values.

x	y
±4	0
±6	≈ ±2.7
±7	≈ ±3.4

65. Joe has 80 coins, all nickels and dimes. The value of the coins is $6.75. How many of each coin does Joe have?

Simpler word form:

$\begin{bmatrix} \text{Number} \\ \text{of} \\ \text{nickels} \end{bmatrix} + \begin{bmatrix} \text{Number} \\ \text{of} \\ \text{dimes} \end{bmatrix} = 80$ coins

$\begin{bmatrix} \text{Value} \\ \text{of} \\ \text{nickels} \end{bmatrix} + \begin{bmatrix} \text{Value} \\ \text{of} \\ \text{dimes} \end{bmatrix} = \$6.75$

Select variables:
Let x represent the number of nickels, so 0.05x represents the value of the nickels.
Let y represent the number of dimes, so 0.10y represents the value of the dimes.

Translate to algebra:

$$x + y = 80 \quad (1)$$
$$0.05x + 0.10y = 6.75 \quad (2)$$

$5x + 10y = 675 \quad (2)$	Multiply equation (2) by 100 to eliminate the decimals.
$\underline{-5x - 5y = -400} \quad (1)$	Multiply equation (1) by -5.
$5y = 275$	
$y = 55$	
$x + 55 = 80 \quad (1)$	Substitute $y = 55$ in equation (1).
$x = 25$	

The solution set is $\{(25, 55)\}$.

Answer:
Joe has 25 nickels and 55 dimes.

EXERCISES 9.4 FINDING LOGARITHMS AND ANTILOGARITHMS

A

Find these logarithms without using tables or calculator:

1. $\log_{10}\frac{1}{10} = \log_{10}10^{-1}$ $\qquad$ $\frac{1}{10} = 10^{-1}$

   $\qquad\qquad = -1(\log_{10}10)$ $\qquad$ Law Three of logarithms.

   $\qquad\qquad = -1$

5. $\log_{10}\frac{1}{1000} = \log_{10}\frac{1}{10^3}$

   $\qquad\qquad = \log_{10}10^{-3}$
   $\qquad\qquad = -3\log_{10}10$
   $\qquad\qquad = -3$

9. $\ln\sqrt{e}$

$x = \ln\sqrt{e}$	Set equal to x.
$e^x = \sqrt{e}$	Write in exponential form.
$e^x = e^{1/2}$	
$x = \frac{1}{2}$	If $b^x = b^y$, then $x = y$.

Find the antilogarithm without using a table or calculator:

13. $\log x = 4$
    $10^4 = x$
    $10,000 = x$

    The base is 10.
    Write in exponential form.

    The antilogarithm is 10,000.

17. $\log x = 6$
    $10^6 = x$
    $1000000 = x$

    The base is 10.
    Write in exponential form.

    The antilogarithm is 1,000,000.

21. $\ln x = 0$
    $10^0 = x$
    $1 = x$

    The base is 10.
    Write in exponential form.

    The antilogarithm is 1.

B

Find the logarithms:

25. $\log 2.3$

    From the table:

    $\log 2.3 \approx 0.3617$

    By calculator:

ENTER	DISPLAY
2.3	2.3
log x	0.361727836

    $\log 2.3 \approx 0.3617$.

29. $\log 8.8$

    From table:
    $\log 8.8 \approx 0.9445$

    By calculator:

ENTER	DISPLAY
8.8	8.8
log x	0.9444826722

    $\log 8.8 \approx 0.9445$

Using a calculator, find the logarithms:

33. ln 16.2

ENTER	DISPLAY
16.2	16.2
ln x	2.785011242

ln 16.2 ≈ 2.7850

Find the antilogarithms:

37. log x = 2.2304

ENTER	DISPLAY
2.2304	2.2304
10^x	169.9808513

The antilogarithm 2.2304 is approximately 169.9809.

41. log x = 0.7267 − 5

ENTER	DISPLAY
0.7267	0.7267
−	0.7267
5	5
=	−4.2733
10^x	5.32966608E − 05 Scientific notation.

≈ 0.000053296608
≈ 0.0001

The antilogarithm (0.7267 − 5) is approximately 0.0001.

C

Find the logarithms:

45. log 230

    ENTER   DISPLAY

     | 230 | 230

     | log x | 2.361727836

log 230 ≈ 2.3617

Find the antilogarithms:

49. log x = 2.6522

    ENTER   DISPLAY

     | 2.6522 | 2.6522

     | 10^x | 448.9520924

The antilogarithm 2.6522 is 448.9521.

53. log x = 4.7777

    ENTER   DISPLAY

     | 4.7777 | 4.7777

     | 10^x | 59937.68983

             ≈ 5.99×10^4 Scientific notation.

The antilogarithm 4.7777 is approximately 59937.6898 or 5.99×10^4 in scientific notation.

Using a calculator, find the following antilogarithms:

57. $\log_e x = 2$

ENTER	DISPLAY
2	2
e^x	7.389056099

The antilogarithm 2, base e, is approximately 7.389.

D

61. The Slippery Rock band plays with an intensity that is 10^{12} times the least intensity that can be heard. What is the decibel level? (Use the formula: $S(i) = 10 \log \frac{i}{i_0}$ where i_0 is the least intensity that can be heard by the human ear and $S(i)$ is measured in decibels.)

Formula:
$$S(i) = 10 \log \frac{i}{i_0}$$

Solve:
$S(i) = 10 \log \dfrac{10^{12} i_0}{i_0}$ Substitute $i = 10^{12}$
$= 10 \log 10^{12}$
$= 12 \cdot 10 \log 10$ Law Three of logarithms.
$= 120 \cdot 1$ $\log 10 = \log_{10} 10 = 1$
$= 120$

Answer:
The Slippery Rock band plays with an intensity of 120 decibels.

65. Find the pH of a mixture with a hydrogen ion concentration of 6.2×10^{-6}. (Use the formula: $pH = \log\frac{1}{H^+}$.)

Formula:
$$pH = \log\frac{1}{H^+}$$

Substitute:
$$pH = \log\frac{1}{6.2 \times 10^{-6}} \qquad H^+ = 6.2 \times 10^{-6}$$

Solve:
$$pH \approx \log 161290.32 \qquad \frac{1}{6.2 \times 10^{-6}} \approx 161290.32$$
$$\approx 5.2 \qquad \text{By calculator.}$$

Answer:
The pH of the mixture is approximately 5.2.

E Maintain Your Skills

69. Find the zeros of $g(x)$ if $g(x) = (x-9)^2 - 121$.

$$\begin{aligned} 0 &= (x-9)^2 - 121 \\ 0 &= x^2 - 18x + 81 - 121 \\ 0 &= x^2 - 18x - 40 \\ 0 &= (x-20)(x+2) \\ x &= 20 \text{ or } x = -2 \end{aligned}$$

Substitute $g(x) = 0$ and solve for x.

Factor.

The zeros of $g(x)$ are 20 and -2.

Write the equation for $G^{-1}(x)$ if:

73. $G(x) = -3x + 6$
$$\begin{aligned} G(x) = y &= -3x + 6 \\ x &= -3y + 6 \qquad \text{Interchange x and y.} \\ 3y &= -x + 6 \qquad \text{Solve for y.} \\ y &= -\tfrac{1}{3}x + 2 \\ G^{-1}(x) &= -\tfrac{1}{3}x + 2 \qquad \text{Replace y with } G^{-1}(x). \end{aligned}$$

77. Two cross-country runners leave the same starting point one hour apart. The first to depart runs 5 mph, and the second runs 7 mph. How far will they have to run when the second overtakes the first?

Simpler word form:
The second runner's time was one hour less than the first runner's time for an equal distance.

Select variable:
Let x represent the distance that the runners had run when the second overtook the first.

	Distance	Rate (mph)	Time = $\dfrac{\text{Distance}}{\text{Rate}}$
First runner	x	5	$\dfrac{x}{5}$
Second runner	x	7	$\dfrac{x}{7}$

Make a table to organize data.

Translate to algebra:
$\dfrac{x}{7} = \dfrac{x}{5} - 1$

Solve:
$5x = 7x - 35$
$-2x = -35$
$x = 17.5$

Multiply both sides by 35 to eliminate the fractions.

Answer:
The runners will run 17.5 miles before the second overtakes the first.

EXERCISES 9.5 EXPONENTIAL EQUATIONS, LOGARITHMIC EQUATIONS, AND FINDING LOGARITHMS, ANY BASE

A

Solve these exponential and logarithmic equations without using tables:

1. $4^x = 32$
   $(2^2)^x = 2^5$
   $2^{2x} = 2^5$
   $2x = 5$
   $x = \dfrac{5}{2}$

Rewrite both sides using 2 as the base.

If $a^x = a^y$, then $x = y$.

5. $\log_3 2 + \log_3 x = 2$
   $\log_3(2x) = 2$ Law One of logarithms.
   $3^2 = 2x$ Write in exponential form.
   $9 = 2x$
   $x = \dfrac{9}{2}$

9. $5^{x+1} = 125^x$
   $5^{x+1} = (5^3)^x$ Rewrite both sides using 5 as a base.
   $5^{x+1} = 5^{3x}$
   $x + 1 = 3x$
   $1 = 2x$
   $x = \dfrac{1}{2}$

13. $4^{2x+1} = 32^{x+5}$
    $2^{2(2x+1)} = 2^{5(x+5)}$ Rewrite both sides using 2 as the base.
    $2(2x + 1) = 5(x + 5)$ If $a^x = a^y$, then $x = y$.
    $4x + 2 = 5x + 25$
    $-23 = x$

B

Use the formulas of this section and a log table or calculator to find each of the logarithms correct to four decimal places:

17. $\log_7 25$

    <u>Formula:</u>
    $$\log_a x = \dfrac{\log_b x}{\log_b a}$$

    <u>Substitute:</u>
    $\log_7 25 = \dfrac{\log_{10} 25}{\log_{10} 7}$ Substitute $x = 25$, $b = 10$, and $a = 7$.

    $\approx \dfrac{1.39794}{0.845098}$ By calculator.

    ≈ 1.1654175

    $\log_7 25 \approx 1.6542$

21. $\log_{13} 6$

Formula:
$$\log_a x = \frac{\log_{10} x}{\log_{10} a}$$

Substitute:
$$\log_{13} 6 = \frac{\log_{10} 6}{\log_{10} 13}$$

$$\approx \frac{0.7781513}{1.1139434} \qquad \text{By calculator.}$$

$$\approx 0.6985555$$

$\log_{13} 6 \approx 0.6986$

Use tables (base 10) or a calculator to solve this exponential equation (round to four decimal places):

25. $\quad 25^x = 50$

$\log 25^x = \log 50$ — Take the logarithm, base 10, of each side.

$x \log 25 = \log 50$ — Property Seven of logarithms.

$x = \dfrac{\log 50}{\log 25}$

$\approx \dfrac{1.6989700}{1.3979400}$ — By calculator.

≈ 1.2153383

So $x \approx 1.2153$.

Solve this logarithmic equation:

29. $\log_2 x - \log_2 (x - 2) = 2$

$\log_2 \dfrac{x}{x-2} = 2$ — Property Six of logarithms.

$2^2 = \dfrac{x}{x-2}$ — Rewrite in exponential form.

$4 = \dfrac{x}{x-2}$

$4(x - 2) = x$ — Multiply both sides by $(x - 2)$.
$4x - 8 = x$
$3x = 8$
$x = \dfrac{8}{3}$

C

Use the formulas of this section and a log table or calculator to find each of the logarithms correct to four decimal places:

33. $\log_{2.5} 45$

$$\log_a x = \frac{\log_b x}{\log_b a}$$ Formula for changing the base of a logarithm.

$$\log_{2.5} 45 = \frac{\log_{10} 45}{\log_{10} 2.5}$$ Substitute 45 for x, 2.5 for a, and 10 for b.

$$\approx \frac{1.6532125}{0.3979400}$$ By calculator.

$$\approx 4.1544265$$

$\log_{2.5} 45 \approx 4.1544$.

37. $\log_{1/5} 0.33$

$$\log_a x = \frac{\log_b x}{\log_b a}$$

$$\log_{1/5} 0.33 = \frac{\log_{10} 0.33}{\log_{10}\left(\frac{1}{5}\right)}$$ $\frac{1}{5} = 0.2$

$$\approx \frac{-0.4814861}{-0.6989700}$$

$$\approx 0.6888508$$

$\log_{1/5} 0.33 \approx 0.6889$

Use tables (base 10) or a calculator to solve this exponential equation (round to four decimal places):

41. $3^{x+1} = 4$

$\log 3^{x+1} = \log 4$ Take the log, base 10, of both sides. Property Seven.
$(x + 1)(\log 3) = \log 4$

$x \log 3 + \log 3 = \log 4$
$x \log 3 = \log 4 - \log 3$
$x = \frac{\log 4 - \log 3}{\log 3}$

$$x \approx \frac{0.6020600 - 0.4771213}{0.4771213}$$ By calculator.

$$\approx \frac{0.1249387}{0.4771213}$$

$$\approx 0.2618595$$

So $x \approx 0.2619$.

Solve these logarithmic equations:

45. $\log_3(x + 6) + \log_3 x = 3$
    $\log_3[x(x + 6)] = 3$ Law One.
    $3^3 = x^2 + 6x$ Write in exponential form.
    $x^2 + 6x - 27 = 0$
    $(x + 9)(x - 3) = 0$
    $x = -9$ or $x = 3$
    $x = 3$ Powers restricted to positive values.

49. $\log_2 x + \log_2(x - 2) = 3$
    $\log_2[x(x - 2)] = 3$ Property Five.
    $2^3 = x(x - 2)$ Write in exponential form.
    $8 = x^2 - 2x$
    $x^2 - 2x - 8 = 0$ Standard form.
    $(x - 4)(x + 2) = 0$
    $x = 4$ or $x = -2$ Reject -2 as $\log x$ is defined only for $x > 0$.
    So $x = 4$.

D

53. How long will it take for $50 to triple in value? (Use the formula in the application.)

    Formula:
    $$V = P\left[1 + \frac{r}{4}\right]^{4y}$$

    Substitute:
    $$150 = 50\left[1 + \frac{0.08}{4}\right]^{4y}$$
    Substitute 50 for P, 8% or 0.08 for r (from Exercise 51) and 150 for V since the value will triple 50.

    $3 = (1 + 0.02)^{4y}$ Divide both sides by 50.
    $\log 3 = 4y \log(1.02)$ Take the log of each side.

    $$4y = \frac{\log 3}{\log 1.02}$$

    $$y = \frac{\log 3}{4\log 1.02}$$

    ≈ 13.86953 Using a calculator.

    It takes approximately 13.87 years for $50 to triple.

E Maintain Your Skills

Evaluate:

57. $\begin{vmatrix} 1 & -1 & -2 \\ -3 & -1 & 2 \\ 1 & 1 & 0 \end{vmatrix}$

Expand by minors. Use the third row since it has a zero.

$= (1)\begin{vmatrix} -1 & -2 \\ -1 & 2 \end{vmatrix} - 1\begin{vmatrix} 1 & -2 \\ -3 & 2 \end{vmatrix} + 0\begin{vmatrix} 1 & -1 \\ -3 & -1 \end{vmatrix}$

$= (-2 - 2) - 1(2 - 6) + 0$
$= -4 + 4$
$= 0$

Solve by any method:

61. $\begin{cases} x^2 + y^2 = 25 & (1) \\ x + y - 1 = 0 & (2) \end{cases}$

$x + y - 1 = 0$
$\quad y = -x + 1$ Solve equation (2) for y.

$x^2 + (-x + 1)^2 = 25$ Substitute $-x + 1$ for y in equation (1).
$x^2 + x^2 - 2x + 1 = 25$
$2x^2 - 2x - 24 = 0$
$x^2 - x - 12 = 0$
$(x - 4)(x + 3) = 0$
$x = 4 \text{ or } x = -3$

$4 + y - 1 = 0$ Substitute each value for x in equation (2).
$\quad y = -3$
$-3 + y - 1 = 0$
$\quad y = 4$

So the solution set is $\{(4 - 3), (-3, 4)\}$.

65. Mr. Mitchell has $10,000 invested, part in a money market fund earning 15% and the rest in a savings account that pays 6%. If the yearly interest from the two investments is $1230, how much is invested in each account?

Simpler word form:

$$\begin{bmatrix} \text{Interest earned} \\ \text{from money} \\ \text{invested at 15\%} \end{bmatrix} + \begin{bmatrix} \text{Interest earned} \\ \text{from money} \\ \text{invested at 6\%} \end{bmatrix} = 1230$$

Select a variable:
Let x represent the amount of money that is invested at 15% interest; then 10000 - x is the amount of money that is invested at 6% interest.

Translate to algebra:
0.15x + 0.06(10000 - x) = 1230

Solve:
0.15x + 600 - 0.06x = 1230
0.09x = 630
x = 7000
and 10000 - x = 10000 - 7000
= 3000

So $7000 is invested at 15% interest, and the rest, $3000 is invested at 6% interest.

CHAPTER 10

FUNCTIONS OF COUNTING NUMBERS

EXERCISES 10.1   SEQUENCES AND SERIES

A

Write the first five terms of the sequence defined by the following:

1. $c(n) = n + 6$
    $c(1) = 1 + 6 = 7$
    $c(2) = 2 + 6 = 8$
    $c(3) = 3 + 6 = 9$
    $c(4) = 4 + 6 = 10$
    $c(5) = 5 + 6 = 11$

    Replace n with 1, 2, 3, 4, and 5. Simplify.

Write a rule for a sequence function given the first five terms:

5. 3, 5, 7, 9, 11
    $2(1) + 1 = 3$
    $2(2) + 1 = 5$
    $2(3) + 1 = 7$
    $2(4) + 1 = 9$
    $2(5) + 1 = 11$

    The terms are consecutive odd numbers. To generate the set of even numbers, the rule would be 2n, so to generate the set of odd numbers, use 2n + 1.

    The sequence rule is $c(n) = 2n + 1$.

9. 3, 6, 9, 12, 15
    $3(1) = 3$
    $3(2) = 6$
    $3(3) = 9$
    $3(4) = 12$
    $3(5) = 15$

    The terms are every third number. To generate the set, multiply n by 3.

    The sequence rule is $c(n) = 3n$.

Write in expanded form, and find the sum:

13. $\sum_{n=1}^{3} (2n + 7)$

    Expand by replacing n with 1, 2, and 3.

    $= [2(1) + 7] + [2(2) + 7] + [2(3) + 7]$
    $= 9 + 11 + 13$
    $= 33$

    Simplify.

17. $\sum_{n=1}^{5} (-2n - 5)$ Expand by replacing n with 1, 2, 3, 4, and 5.

$= [-2(1) - 5] + [-2(2) - 5] + [-2(3) - 5]$
$\quad + [-2(4) - 5] + [-2(5) - 5]$
$= (-7) + (-9) + (-11) + (-13) + (-15)$
$= -55$

Write a rule for a sequence function given the first five terms and find c(8):

21. $1, \frac{1}{4}, \frac{1}{9}, \frac{1}{16}, \frac{1}{25}$

$\frac{1}{1^2} = 1$ The terms are the reciprocals of the squares of consecutive counting numbers.

$\frac{1}{2^2} = \frac{1}{4}$

$\frac{1}{3^2} = \frac{1}{9}$

$\frac{1}{3^2} = \frac{1}{16}$

$\frac{1}{4^2} = \frac{1}{16}$

$\frac{1}{5^2} = \frac{1}{25}$

$c(n) = \frac{1}{n^2}$

$c(8) = \frac{1}{8^2} = \frac{1}{64}$

Write in expanded form, and find the sum:

25. $\sum_{n=1}^{6} (4n - n^2)$

$= (4 \cdot 1 - 1^2) + (4 \cdot 2 - 2^2) + (4 \cdot 3 - 3^2)$
$\quad + (4 \cdot 4 - 4^2) + (4 \cdot 5 - 5^2) + (4 \cdot 6 - 6^2)$
$= 3 + 4 + 3 + 0 + (-5) + (-12)$
$= 10 - 17$
$= -7$

29. $\sum_{n=1}^{9} (n^3 - n^2)$

$= (1^3 - 1^2) + (2^3 - 2^2) + (3^3 - 3^2) + (4^3 - 4^2) + (5^3 - 5^2)$
$\quad + (6^3 - 6^2) + (7^3 - 7^2) + (8^3 - 8^2) + (9^3 - 9^2)$
$= 0 + 4 + 18 + 48 + 100 + 180 + 294 + 448 + 648$
$= 1740$

Write the following series in summation notation:

33. $1 + \frac{1}{8} + \frac{1}{27} + \frac{1}{64} + \frac{1}{125} + \ldots$ $1 = \frac{1}{1}$. The denominators are cubes of successive counting numbers.

$c(n) = \frac{1}{n^3}$

$\sum_{n=1}^{\infty} \frac{1}{n^3}$ The ellipsis (three dots) indicates there is no boundary for n.

C

Write a rule for a sequence function given the first five terms and find c(9). Write the summation notation for the first 10 terms.

37. $\frac{1}{7}, \frac{2}{8}, \frac{3}{9}, \frac{4}{10}, \frac{5}{11}$ The numerators are successive counting numbers. The denominators are respectively six more than the numerators.

$c(n) = \frac{n}{n+6}$

$c(9) = \frac{9}{9+6} = \frac{9}{15} = \frac{3}{5}$ Replace n with 9, and reduce.

$\sum_{n=1}^{10} \frac{n}{n+6}$

41. 0, 2 log 2, 3 log 3, 4 log 4, 5 log 5

$c(n) = n \log n$ $0 = 1 \log 1$ since $\log 1 = 0$.
$c(9) = 9 \log 9$

$\sum_{n=4}^{10} n \log n$

45. $5x^3$, $4x^6$, $3x^9$, $2x^{12}$, x^{15}

$$c(n) = (6 - n)x^{3n}$$

$$c(9) = (6 - 9)x^{3 \cdot 9}$$
$$= -3x^{27}$$

$$\sum_{n=1}^{10} (6 - n)x^{3n}$$

The coefficients of the x terms are expressed 6 - n. The powers on the x terms are multiples of 3.
Replace n with 9, and simplify.

Write in expanded form, and find the sum:

49. $\sum_{n=4}^{7} (n^2 - 5n)$

$$= (4^2 - 5 \cdot 4) + (5^2 - 5 \cdot 5)$$
$$\quad + (6^2 - 5 \cdot 6) + (7^2 - 5 \cdot 7)$$
$$= (-4) + 0 + 6 + 14$$
$$= 16$$

Replace n with 4, 5, 6, and 7 successively and simplify.

53. $\sum_{n=10}^{14} (-2n + 5)$

$$= (-2 \cdot 10 + 5) + (-2 \cdot 11 + 5) + (-2 \cdot 12 + 5) + (-2 \cdot 13 + 5)$$
$$\quad + (-2 \cdot 14 + 5)$$
$$= (-15) + (-17) + (-19) + (-21) + (-23)$$
$$= -95$$

Replace n with 10, 11, 12, 13, and 14 successively and simplify.

D

57. For every year that strawberries are planted in the same field, the yield is reduced by 8 tons. If the yield for the first year is 91 tons, what is the yield in the fourth year?

$$c(n) = 91 - 8(n - 1)$$

$$c(4) = 91 - 8(4 - 1)$$
$$= 91 - 8(3) = 91 - 24$$
$$= 67$$

Write the rule for the yearly yield.
Replace n with 4.
Simplify.

The yield in the fourth year is 67 tons.

61. A rubber ball dropped from 10 ft rebounds 5 ft on the first bounce and 2.5 ft on the second bounce. If it continues to rebound each time to one-half of the previous height, how high will it rebound on the eighth bounce?

5, 2.5, 1.25, 0.625, 0.3125, 0.15625, 0.078125, 0.0390625

The rebound is one-half the previous height, so write the first eight terms beginning at 5, the first rebound.

Alternative solution:

$c(n) = \dfrac{5}{2^{n-1}}$ Write the rule for the sequence.

$c(8) = \dfrac{5}{2^{8-1}} = \dfrac{5}{2^7}$ Replace n with 8, and simplify.

$= \dfrac{5}{128}$

<u>Answer:</u>
The ball will rebound 0.0390625 ft of $\dfrac{5}{128}$ ft on the eighth bounce.

E. Maintain Your Skills

65. Write the coordinate of the y-intercept of the graph of $3x + 5y = 8$.

$3(0) + 5y = 8$ Replace x with 0 and solve for y.
$5y = 8$
$y = \dfrac{8}{5}$

The y-intercept is $\left(0, \dfrac{8}{5}\right)$.

69. Write the coordinates of the x-intercepts of the graph of $y = 2(x-7)^2 + 10$.

$0 = 2(x - 7)^2 + 10$ Replace y withh 0, and solve for x.
$0 = 2(x^2 - 14x + 49) + 10$
$0 = 2x^2 - 28x + 98 + 10$
$0 = 2x^2 - 28x + 108$
$0 = x^2 - 14x + 54$ Divide both sides by 2.

$x = \dfrac{14 \pm \sqrt{196 - 216}}{2}$ Use the quadratic formula to solve for x.

$x = \dfrac{14 \pm \sqrt{-20}}{2}$

There are no x-intercepts, since $\sqrt{-20}$ is not a real number.

EXERCISES 10.2 ARITHMETIC PROGRESSIONS (SEQUENCES)

A

Write the first five terms of the arithmetic progression and find the indicated term:

1. $a_1 = 7$, $d = 3$. Find a_{10}

$a_n = a_1 + (n - 1)d$ Formula for the nth term of an arithmetic progression.
$a_1 = 7$
$a_2 = 7 + (2 - 1)3 = 10$ Substitute $n = 2, 3, 4$, and 5 in the formula.
$a_3 = 7 + (3 - 1)3 = 13$
$a_4 = 7 + (4 - 1)3 = 16$
$a_5 = 7 + (5 - 1)3 = 19$
$a_{10} = 7 + (10 - 1)3 = 34$ Substitute $a = 10$ in the formula.

5. $a_1 = 7$, $a_2 = 19$. Find a_{11}
   $d = 12$

     Find d, the difference between two successive terms. $19 - 7 = 12$.

$a_n = a + (n - 1)d$ Formula.
$a_1 = 7$ First term.
$a_2 = 19$ Given.
$a_3 = 7 + (3 - 1)12 = 7 + 24 = 31$ Substitute 3, 4, and 5.
$a_4 = 7 + (4 - 1)12 = 7 + 36 = 43$
$a_5 = 7 + (5 - 1)12 = 7 + 48 = 55$
$a_{11} = 7 + (11 - 1)12 = 127$ Substitute $n = 11$.

9. $a_1 = 2$, $a_4 = 14$. Find a_6
   $n = 1$, $a_1 = 2$
   $n = 4$, $a_4 = 14$ Given information.

$a_n = a_1 + (n - 1)d$ Formula for general term.
$14 = 2 + (4 - 1)d$ Substitute 14 for a_n, 2 for a_1 and 4 for n.
$14 = 2 + 3d$
$12 = 3d$
$4 = d$ Solve for d.

$a_1 = 2$
$a_2 = 2 + 1(4) = 6$
$a_3 = 2 + 2(4) = 10$
$a_4 = 2 + 3(4) = 14$
$a_5 = 2 + 4(4) = 18$
$a_6 = 2 + 5(4) = 22$

Find the number of terms in each of the following finite arithmetic progressions:

13. 17, 13, ..., -23
    $a_1 = 17$; $d = 13 - 17 = -4$
    $a_n = -23$

$a_n = a_1 + (n - 1)d$	Formula for n^{th} term.
$-23 = 17 + (n - 1)(-4)$	Substitue $a_n = -23$, $a_1 = 17$,
$-23 = 17 - 4n + 4$	and $d = -4$.
$-23 = 21 - 4n$	Solve for n.
$-44 = -4n$	
$11 = n$	

B

Find all the terms between the given terms of the following arithmetic progressions:

17. $a_1 = 10$, $a_6 = 0$

$a_n = a_1 + (n - 1)d$	Formula.
$0 = 10 + (6 - 1)d$	Substitute 0 for a_n, 10 for
$0 = 10 + 5d$	a_1, and 6 for n.
$-10 = 5d$	
$-2 = d$	Solve for d.
$a_1 = 10$	Generate the following terms
$a_2 = 10 - 2 = 8$	by subtracting 2.
$a_2 = 8 - 2 = 6$	
$a_4 = 6 - 2 = 4$	
$a_5 = 4 - 2 = 2$	

21. $a_{10} = 22$, $l_{15} = 37$

$n = 10$, $a_{10} = 22$	Given information.
$n = 15$, $a_{15} = 37$	
$a_n = a_1 + (n - 1)d$	
	Write two equations with two variables to find a_1 and d.
$22 = a_1 + 9d$ (1)	Substitute $n = 10$ and $a_{10} = 22$ to get equation (1).
$37 = a_1 + 14d$ (2)	Substitute $n = 15$ and $a_{15} = 37$ to get equation (2).
$15 = 5d$	Solve the system by subtracting (1) from (2).
$3 = d$	

$a_{10} = 22$
$a_{11} = 25$
$a_{12} = 28$
$a_{13} = 31$
$a_{14} = 34$

Start with a_{10} and add 3 to get the terms that are between a_{10} and a_{15}.

So the missing terms are 25, 28, 31, and 34.

Find the sum of the indicated number of terms of the arithmetic progressions.

25. 15, 9, 3, ... 10 terms.
$a_1 = 5$, $d = -6$, $n = 10$

$a_{10} = 15 + (10 - 1)(-6)$
$= 15 + 9(-6)$
$= -39$

Write the tenth term.

$S_n = \dfrac{n(a_1 + a_n)}{2}$

Formula for the sum of n terms of an arithmetic progression.
Substitute $n = 10$, $a_1 = 15$, and $a_n = -39$.

$= \dfrac{10[15 + (-39)]}{2}$

$= 5(-24)$
$= -120$

29. $-2, -\dfrac{3}{2}, -1, \ldots$ 9 terms
$a_1 = -2$, $d = \dfrac{1}{2}$, $n = 9$

$a_9 = -2 + (9 - 1)\dfrac{1}{2}$

Write the ninth term.
$d = -\dfrac{3}{2} + 2 = \dfrac{1}{2}$.

$= -2 + 8\left[\dfrac{1}{2}\right]$
$= -2 + 4$
$= 2$

$S_n = \dfrac{n(a_1 + a_n)}{2}$

$= \dfrac{9(-2 + 2)}{2}$

$= 0$

Write the first five terms of the arithmetic progression and find the indicated term:

33. $a_1 = 5$, $a_2 = 3.5$. Find a_{15}.
     $d = 3.5 - 5 = -1.5$

   $a_1 = 5$
   $a_2 = 5 - 1.5 = 3.5$
   $a_3 = 3.5 - 1.5 = 2$
   $a_4 = 2 - 1.5 = 0.5$
   $a_5 = 0.5 - 1.5 = -1$

   $a_n = a_1 + (n - 1)d$

   $a_{15} = 5 + (15 - 1)(-1.5)$
   $\phantom{a_{15}} = 5 + 14(-1.5)$
   $\phantom{a_{15}} = 5 + (-21)$
   $\phantom{a_{15}} = -16$

   So $a_{15} = -16$.

37. $a_5 = 4$, $a_{15} = -11$. Find a_{31}.

   $a_n = a_1 + (n - 1)d$ Find a_1 and d by writing two equations in two variables, a_1 and d.

   $4 = a_1 + 4d$ (1)   Substitute 4 for a_5 and 5 for n for equation (1).

   $-11 = a_1 + 14d$ (2)   Substitute -11 for a_{15} and 15 for n for equation (2).
   $-15 = 10d$
   $-\frac{3}{2} = d$ Subtract equation (1) from equation (2).

   $4 = a_1 + 4\left[-\frac{3}{2}\right]$ Substitute $d = -\frac{3}{2}$ in
   $ = a_1 - 6$ equation (1) and solve for a_1.

   $10 = a_1$

   $a_{31} = a_1 + (n - 1)d$ Formula for general term.
   $\phantom{a_{31}} = 10 + (31 - 1)\left[-\frac{3}{2}\right]$ Substitute 31 for n and $-\frac{3}{2}$ for d, 10 for a_1.

   $\phantom{a_{31}} = 10 + 30\left[-\frac{3}{2}\right]$
   $\phantom{a_{31}} = 10 - 45$
   $\phantom{a_{31}} = -35$

C

Find the number of terms and the sum of the following finite arithmetic progression.

41. 14.2, 12.9, ..., -14.4
$a_1 = 144.2$, $d = 12.9 - 14.2 = -1.3$, $a_n = -14.4$

$a_n = a_1 + (n - 1)d$
$-14.4 = 14.2 + (n - 1)(-1.3)$ Substitute for a_n, a_1, and d.
$-14.4 = 14.2 - 1.3n + 1.3$ Solve for n.
$-14.4 = 15.5 - 1.3n$
$-29.9 = -1.3n$
$23 = n$

$S_n = \dfrac{n(a_1 + a_n)}{2}$

$= \dfrac{23[14.2 + (-14.4)]}{2}$

$= \dfrac{23(-0.2)}{2}$

$= -2.3$

The number of terms is 23, and the sum of the terms is -2.3.

Find the indicated sum:

45. $\sum\limits_{n=1}^{18} (3n + 9)$

$a_1 = 3(1) + 9 = 12$ Generate the first and last
$a_{18} = 3(18) + 9 = 63$ terms by using the rule
$n = 18$ given.

$S_n = \dfrac{n(a_1 + a_{18})}{2}$ Formula for the sum of an arithmetic progression.

$= \dfrac{18(12 + 63)}{2}$

$= 9(75)$

$= 675$

49. $\sum_{n=1}^{25} (-0.4n - 10)$

$a_1 = -0.4(1) - 10 = -10.4$ Generate the first and last
$a_{25} = -0.4(25) - 10 = -20$ terms by using the rule
$n = 25$ given.

$S_n = \dfrac{n(a_1 + a_{25})}{2}$ Formula for the sum of an arithmetic progression.

$= \dfrac{25[-10.4 + (-20)]}{2}$ Substitute $n = 25$, $a_1 = -10.4$ and $a_{25} = -20$.

$= \dfrac{25(-30.4)}{2}$

$= -380$

Find the sum of the indicated number of terms of the following progression:

53. $a_5 = -5$, $a_{15} = -30$. Find S_{36}.
$a_n = a_1 + (n - 1)d$

$-5 = a_1 + 4d$ (1) First find a_1 and d by
$-30 = a_1 + 14d$ (2) generating a system of two equations in two variables.

$-25 = 10d$
$-2.5 = d$ Subtract (1) from (2).
$-5 = a_1 + 4(-2.5)$ Solve for a_1 using
$-5 = a_1 - 10$ equation (1).
$5 = a_1$

$a_n = a_1 + (n - 1)d$ Formula for general term.
$a_{36} = 5 + 35(-2.5)$ Substitue $d = -2.5$ and $a_1 = 5$.
$= -82.5$

$S_n = \dfrac{36[5 + (-82.5)]}{2}$ Formula for the sum of an arithmetic progression.
$= 18(-77.5)$
$= -1395$

The sum of the first 36 terms is -1395.

D

57. Pete intends to increase the distance he runs each week by two miles. If he now runs 15 miles per week, in how many weeks will he be running 3 miles per week?

$15, 17, 19, \ldots 63$ — Write the first few terms of the progression.

$a_1 = 15,\ d = 17 - 15 = 2,\ a_n = 63$

$a_n = a_1 + (n - 1)d$ — Formula for the nth term.
$63 = 15 + (n - 1)2$ — Substitute and solve for n.
$63 = 15 + 2n - 2$
$63 = 13 + 2n$
$50 = 2n$
$25 = n$

In 25 weeks, Pete will be running 63 miles per week.

61. Willy is stacking blocks. How many blocks are there in his stack if there are 18 blocks in the bottom row, 17 in the second row, 16 in the third row, and so on, until there is 1 block in the top row?

$S_n = 18 + 17 + 16 + \ldots + 1$ $a = 18,\ a_n = 1$
    $d = 17 - 18 = -1$
$1 = 18 + (n - 1)(-1)$ Substitute into the formula
$1 = 18 - n + 1$ for the general term to find
$1 = 19 - n$ n.
$n = 18$

$S_{18} = \dfrac{18(18 + 1)}{2} = 171$

There are 171 blocks.

E Maintain Your Skills

Write in exponential form:

65. $\log_7 49 = 2$
    $7^2 = 49$

7 is the base, 2 the exponent, and 49 is the power.

Write in logarithmic form:

69. $4096^{1/6} = 4$
    $\log_{4096} 4 = \dfrac{1}{6}$

4096 is the base, $\dfrac{1}{6}$ the exponent or logarithm, and 4 the power.

EXERCISES 10.3   GEOMETRIC PROGRESSIONS (SEQUENCES)

A

Write the first four terms of the geometric progression and find the indicated term:

1. $a_1 = 1$, $r = 3$. Find a_6.

   $a_1 = 1$

   $a_2 = 1 \cdot 3 = 3$

   $a_3 = 3 \cdot 3 = 9$

   $a_4 = 9 \cdot 3 = 27$

   The first term is 1, and each term after is found by multiplying the preceding term by 3.

   The first four terms are 1, 3, 9, and 27.

   $a_n = a\, r^{n-1}$

   $a_6 = 1 \cdot 3^{6-1} = 3^5 = 243$

   Formula for the general term of a geometric progression. Substitute $a_1 = 1$, $n = 6$, and $r = 3$.

   The sixth term is 243.

Find the number of terms in each finite geometric progression:

5. 6, 12, ..., 1536

   $r = \dfrac{12}{6} = 2$ Find r.

   $a_n = a_1 r^{n-1}$

   Use the formula to solve for n, where $a_1 = 6$, $r = 2$, and $a_n = 1536$.

   $1536 = 6 \cdot 2^{n-1}$

   $256 = 2^{n-1}$

   $2_8 = 2^{n-1}$

   $n - 1 = 8$

   $n = 9$

   Write each side with a base of 2.

   If $a^x = a^y$, then $x = y$.

   There are 9 terms.

347

Find all the terms between the given terms of the following geometric progressions:

9. $a_1 = 2$, $a_5 = 162$

$a_n = a_1 r^{n-1}$ Formula for the nth term of a geometric progression.

$162 = 2 \cdot r^4$ Substitue 162 for a_n, and 5
$81 = r^4$ for n. Solve for r.
$\pm 3 = r$

$a_2 = 2(\pm 3)^1 = \pm 6$ Use the formula for the nth term to find the terms between the first and fifth terms.
$a_3 = 2(\pm 3)^2 = 18$
$a_4 = 2(\pm 3)^3 = \pm 54$

13. $a_2 = -1$, $a_7 = 243$

$a_n = a_1 r^{n-1}$ Formula for the nth term.

So,
$\begin{cases} -1 = a_1 r^1 & (1) \\ 243 = a_1 r^6 & (2) \end{cases}$ Write two equations with two unknowns using the data to find a_1 and r.

$-\frac{1}{r} = a_1$ Solve (a) for a_1.

$243 = -\frac{1}{r} \cdot r^6$ Substitute in (2).
$243 = -r^5$
$-243 = r^5$
$-3 = 4$

$a_2 = -1$ Use -3 as the common ratio.
$a_3 = -1 \cdot -3 = 3$
$a_4 = 3 \cdot -3 = -9$
$a_5 = -9 \cdot -3 = 27$
$a_6 = 27 \cdot -3 = -81$

Find the sum of the indicated number of terms of the following geometric progression:

17. 2, -6, 18, ... 6 terms
$a_1 = 2$, $r = \dfrac{-6}{2} = -3$

$$S_6 = \dfrac{2 - 2(-3)^6}{1 - (-3)}$$
$$= \dfrac{2 - 1458}{4}$$
$$= -\dfrac{1456}{4}$$
$$= -364$$

The sum of the first 6 terms of the given geometric progression is -364.

Write the first four terms of the geometric progression and find the indicated term:

21. $a_1 = 9$, $a_2 = 6$. Find a_8.
$r = \dfrac{6}{9} = \dfrac{2}{3}$

$a_n = a_1 r^{n-1}$ Formula for the nth term.

$a_1 = 9\left(\dfrac{2}{3}\right)^0 = 9$

$a_2 = 9\left(\dfrac{2}{3}\right)^1 = 6$

$a_3 = 9\left(\dfrac{2}{3}\right)^2 = 4$

$a_4 = 9\left(\dfrac{2}{3}\right)^3 = \dfrac{8}{3} = 2\dfrac{2}{3}$

$a_8 = 9\left(\dfrac{2}{3}\right)^4 = \dfrac{128}{243}$

25. $a_2 = \frac{1}{6}$, $a_5 = \frac{1}{162}$. Find a_7

$a_n = a_1 r^{n-1}$ Formula for the nth term of a geometric progression.

So,
$$\begin{cases} \frac{1}{6} = a_1 r^1 & (1) \\ \frac{1}{162} = a_1 r^4 & (2) \end{cases}$$

Write two equations with two unknowns using the given data to find a_1 and r.

$\frac{1}{6r} = a_1$ Solve (1) for a_1.

$\frac{1}{162} = \frac{1}{6r} \cdot r^4$ Substitute in (2).

$\frac{1}{162} = \frac{r^3}{6}$

$\frac{6}{162} = r^3$ Multiply both sides by 6.

$\frac{1}{27} = r^3$ Reduce the fraction.

$\frac{1}{3} = r$

$\frac{1}{6} = a_1 \left(\frac{1}{3}\right)$ Substitute $\frac{1}{3}$ for r in (1).

$\frac{1}{6} = \frac{a_1}{3}$

$\frac{3}{6} = a_1$ Multiply both sides by 3.

$\frac{1}{2} = a_1$

$a_1 = \frac{1}{2}$

$a_2 = \frac{1}{2}\left(\frac{1}{3}\right)^1 = \frac{1}{6}$ Use the formula for the nth substituting $\frac{1}{2}$ for a_1 and $\frac{1}{3}$ for r.

$a_3 = \frac{1}{2}\left(\frac{1}{3}\right)^2 = \frac{1}{18}$

$a_4 = \frac{1}{2}\left(\frac{1}{3}\right)^3 = \frac{1}{54}$

$a_7 = \frac{1}{2}\left(\frac{1}{3}\right)^6 = \frac{1}{1458}$

Find the number of terms and the sum of the following finite geometric progressions:

29. $80, 20, \ldots, \dfrac{5}{64}$

$a_1 = 80, \quad r = \dfrac{20}{80} = \dfrac{1}{4}, \quad a_n = \dfrac{5}{64}$

$80, 20, 5, \dfrac{5}{4}, \dfrac{5}{16}, \dfrac{5}{64}$ Use the ratio to find the missing terms.

There are 6 terms.

$S_n = \dfrac{a_1 - a_1 r^n}{1 - r}$

$S_6 = \dfrac{80 - 80\left(\dfrac{1}{4}\right)^6}{1 - \dfrac{1}{4}} = \dfrac{80 - \dfrac{1}{4096}}{\dfrac{3}{4}}$

$= 106\dfrac{41}{64}$ or 106.640625

The sum of the six terms is $106\dfrac{41}{64}$.

33. $-2.1, 1.05, \ldots, -0.13125$

$a_1 = -2.1, \quad r = \dfrac{1.05}{-2.1} = -0.5, \quad a_n = -0.13125$

$-2.1, 1.05, -0.525, 0.2625, -0.13125$ Use the ratio to find the missing terms.

There are five terms.

$S_n = \dfrac{a_1 - a_1 r^n}{1 - r}$

$S_5 = \dfrac{-2.1 - (-2.1)(-0.5)^5}{1 - (-0.5)}$ Substitute -2.1 for a_1, and -0.5 for r.

$= \dfrac{-2.1 - (0.065625)}{1.5}$

$= \dfrac{-2.165625}{1.5}$

$= -1.44375$

C

Find the indicated sum:

37. $\sum_{n=1}^{10} 10(-1)^n$

$= 10(-1)^1 + 10(-1)^2 + 10(-1)^3 + 10(-1)^4 + 10(-1)^5 + 10(-1)^6 + 10(-1)^7 + 10(-1)^8 + 10(-1)^9 + 10(-1)^{10}$

$= -10 + 10 - 10 + 10 - 10 + 10 - 10 + 10 - 10 + 10$

$= 0$

41. $\sum_{n=1}^{10} 2187(3)^{-n}$

$n = 11$ Find a_1, r, and n.
$a_1 = 2187(3)^{-1} = 729$
$a_2 = 2187(3)^{-2} = 243$
$r = \dfrac{243}{729} = \dfrac{1}{3}$

$S_n = \dfrac{a_1 - a_1 r^n}{1 - r}$ Substitute.

$S_{11} = \dfrac{729 - 729\left(\dfrac{1}{3}\right)^{11}}{1 - \dfrac{1}{3}} = \dfrac{729 - \dfrac{729}{177147}}{\dfrac{2}{3}}$

$= 1093\dfrac{40}{81}$ or approximately 1093.4938

The sum is $1093\dfrac{40}{81}$.

45. $\sum_{n=1}^{5} (3)\left(-\dfrac{1}{3}\right)^n$

$n = 5$ Find n, a_1, and r.
$a_1 = (3)\left(-\dfrac{1}{3}\right)^1 = -1$
$a_2 = (3)\left(-\dfrac{1}{3}\right)^2 = \dfrac{1}{3}$
$r = \dfrac{a_2}{a_1} = \dfrac{\frac{1}{3}}{-1} = -\dfrac{1}{3}$

So

$$S_n = \frac{a_1 - a_1 r^n}{1 - 4}$$

$$S_5 = \frac{-1 - (-1)\left(-\frac{1}{3}\right)^5}{1 - \left(-\frac{1}{3}\right)}$$ Substitute.

$$= \frac{-\frac{244}{243}}{\frac{4}{3}}$$

$$= -\frac{61}{81}$$

The sum is $-\frac{61}{81}$ or approximately -0.7531.

49. $\sum_{n=1}^{5} \left(\frac{3}{2}\right)\left(\frac{1}{3}\right)^n$

  $n = 5$ Find n, a_1, and r.

$$a_1 = \left(\frac{3}{2}\right)\left(\frac{1}{3}\right)^1 = \frac{1}{2}$$

$$a_2 = \left(\frac{3}{2}\right)\left(\frac{1}{3}\right)^2 = \frac{1}{6}$$

$$r = \frac{a_2}{a_1} = \frac{\frac{1}{6}}{\frac{1}{2}} = \frac{1}{3}$$

$$S_n = \frac{a_1 - a_1 r^n}{1 - r}$$

$$S_5 = \frac{\frac{1}{2} - \frac{1}{2}\left(\frac{1}{3}\right)^5}{1 - \frac{1}{3}}$$ Substitute.

$$= \frac{\frac{242}{486}}{\frac{2}{3}}$$

$$= \frac{121}{162}$$

The sum is $\frac{121}{162}$ or approximately 0.7469.

D

53. The price of a can of juice increases at a rate of 5% per year because of inflation. If the price of the juice is now 73¢, what will the price be four years from now? Twenty years from now? (Hint: Use r = 1.05 since the price each year is 105% of last year's price.)

$a_n = a_1 r^{n-1}$

$a_5 = 0.73(1.05)^5$

≈ 0.89

$a_{21} = 0.73(1.05)^{20}$

≈ 1.94

The progression of the price of the juice in each successive year is geometric. Formula for the nth term of a geometric progression. Substitute a_1 = 0.73 and r = 1.05. Use n = 5 since the fourth year represents a_5 or four additional years after a_1.

Substitute a_1 = 0.73, 4 = 1.05, and n = 21 for the twentieth year.

The price of the juice four years from now will be $0.89 and twenty years from now will be $1.94.

E Maintain Your Skills

Solve without a calculator:

57. $s = \log_{10} \sqrt[3]{10}$
    $= \log_{10} 10^{1/3}$
    $= \frac{1}{3} \log_{10} 10$
    $= \frac{1}{3}$

$\sqrt[3]{10} = 10^{1/3}$

Property Seven of logarithms.

$\log_{10} 10 = 1$

Solve with the help of a calculator or logarithm tables; write the answer to the nearest ten-thousandth:

61. $\log_5 17 = x$

$\dfrac{\log 17}{\log 5} = x$ Change of base formula.

ENTER	DISPLAY
17	17
log	1.2304489
÷	
5	5
log	0.6989700
=	1.7603744

$\log_5 17 \approx 1.7604$

EXERCISES 10.4 INFINITE GEOMETRIC PROGRESSIONS

A

Find the sum of each of the following infinite geometric series:

1. $8 + 4 + 2 + \ldots$

$a_1 = 8,\ r = \dfrac{4}{8} = \dfrac{1}{2}$ Identify a_1 and r.

$S_\infty = \dfrac{a_1}{1 - 4}$

$= \dfrac{8}{1 - \dfrac{1}{2}}$ Substitute 8 for a_1 and $\dfrac{1}{2}$ for r.

$= \dfrac{8}{\dfrac{1}{2}}$

$= 16$

355

5. $21 + 7 + \frac{7}{3} + \ldots$

   $a_1 = 21, \; r = \frac{1}{3}$

   $S_\infty = \frac{21}{\frac{2}{3}} = 31\frac{1}{2}$

9. $30 + 15 + \frac{15}{2} + \ldots$

   $a_1 = 30, \; r = \frac{1}{2}$

   $S_\infty = \frac{30}{\frac{1}{2}} = 60$

13. $-100 + (-50) + (-25) + \ldots$

    $a_1 = -100$

    $r = \frac{-50}{-100} = \frac{1}{2}$

    $S_\infty = \frac{-100}{1 - \frac{1}{2}}$

    $ = -200$

B

17. $12 - 6 + 3 = \ldots$

    $a_1 = 12, \; r = \frac{-6}{12} = -\frac{1}{2}$ Identify a_1 and r.

    $S_\infty = \frac{12}{a - \left[-\frac{1}{2}\right]}$

    $ = \frac{12}{\frac{3}{2}}$

    $ = 8$

21. $\frac{2}{7} - \frac{1}{7} + \frac{1}{14} - \ldots$

$$a_1 = \frac{2}{7} \qquad r = \frac{-\frac{1}{7}}{\frac{2}{7}} = -\frac{1}{2}$$

$$S_\infty = \frac{\frac{2}{7}}{1 - \left(-\frac{1}{2}\right)}$$

$$= \frac{\frac{2}{7}}{\frac{3}{2}}$$

$$= \frac{4}{21}$$

25. $28 - 12.6 + 5.67 = \ldots$

$$a_1 = 28 \qquad r = \frac{-12.6}{28} = -\frac{9}{20}$$

$$S_\infty = \frac{28}{1 - \left(-\frac{9}{20}\right)} = \frac{28}{\frac{29}{20}}$$

$$= \frac{560}{29}$$

$$= 19\frac{9}{29}$$

C

Find the common fraction name for the following repeating decimals:

29. $0.55555 \ldots$

$\qquad = 0.5 + 0.05 + 0.005 + \ldots$ Write as a series to identify a_1 and r.

$$a_1 = 0.5 \qquad r = \frac{0.05}{0.5} = 0.1$$

$$S_\infty = \frac{0.5}{1 - (-0.1)} = \frac{0.5}{0.9} = \frac{5}{9}$$

The common fraction name for $0.55555 \ldots$ is $\frac{5}{9}$.

33. 0.016016016 ...

= 0.016 + 0.000016 + 0.000000016 + ...

$a_1 = 0.016 \quad R = \dfrac{0.000016}{0.016} = 0.001$

$S_\infty = \dfrac{0.016}{1 - 0.001} = \dfrac{0.016}{0.999} = \dfrac{16}{999}$

The common fraction name for 0.016016016 ... is $\dfrac{16}{999}$.

37. 0.108108108 ...

= 0.108 + 0.000108 + 0.000000108 + ...

$a_1 = 0.108 \quad r = \dfrac{0.000108}{0.108} = 0.001$

$S_\infty = \dfrac{0.108}{1 - 0.001} = \dfrac{0.108}{0.999} = \dfrac{4(0.027)}{37(0.027)} = \dfrac{4}{37}$

The common fraction name for 0.108108108 ... is $\dfrac{4}{37}$.

D

41. A mine in Idaho yields 150 tons of high-grade gold ore in the first year and each year thereafter $\dfrac{3}{4}$ the tonnage of the previous year. Find the amount of high-grade ore that was in the mine.

$a_1 = 150, \ r = \dfrac{3}{4}$

$S_\infty = \dfrac{150}{\dfrac{1}{4}} = 600$

There were 600 tons of high-grade ore in the mine.

E Maintain Your Skills

Write as a combination of logarithms of single variables:

45. $\log \dfrac{r^3}{s} = \log r^3 - \log s$ Property Six of logarithms.

       $= 3 \log r - \log s$ Property Seven.

Write as a single logarithm:

49. $2 \log z - 3 \log w = \log z^2 - \log w^3$ Property Seven of logarithms.

$ = \log \dfrac{z^2}{w^3}$ Property Six.

Solve:

53. $\log_5(x + 3) - \log_5 x = 2$

$\log_5\left[\dfrac{x + 3}{x}\right] = 2$ Property Seven of logarithms.

$5^2 = \dfrac{x + 3}{x}$ Write in exponential form.

$25x = x + 3$ Multiply both sides by x.
$24x = 3$
$x = \dfrac{1}{8}$

EXERCISES 10.5   BINOMIAL EXPANSION

A

Evaluate:

1. $3! = 3 \cdot 2 \cdot 1 = 6$ Write in expanded form and simplify.

5. $\dfrac{9!}{6!} = \dfrac{9 \cdot 8 \cdot 7 \cdot 6!}{6!} = 9 \cdot 8 \cdot 7 = 504$

Write in expanded form:

9. $(x + 1)^4$
$n = 4, a = x, b = 1$ Identify a, b, and n.

$\dfrac{4!}{4!0!}(x)^4(1)^0 + \dfrac{4!}{3!1!}(x)(1)^1$ Use the formula to expand.

$+ \dfrac{4!}{2!2!}(x)^2(1)^2 + \dfrac{4!}{1!3!}(x)^1(1)^3$

$+ \dfrac{4!}{0!4!}(x)^0(1)^4$

$= x^4 + 4x^3 + 6x^2 + 4x + 1$ Simplify.

13. $(x + 2)^8$
    $n = 8, a = x, b = 2$ Identify a, b, and n.

$\dfrac{8!}{8!0!}(x)^8(2)^0 + \dfrac{8!}{7!1!}(x)^7(2)^1$ Use the formula to expand $(x + 2)$.

$+ \dfrac{8!}{6!2!}(x)^6(2)^2 + \dfrac{8!}{5!3!}(x)^5(2)^3$

$+ \dfrac{8!}{4!4!}(x)^4(2)^4 + \dfrac{8!}{3!5!}(x)^3(2)^5$

$+ \dfrac{8!}{2!6!}(x)^2(2)^6 + \dfrac{8!}{1!7!}(x)^1(2)^7$

$+ \dfrac{8!}{0!8!}(x)^0(2)^8$

$= (1)(x^8)(1) + (8)(x^7)(2) + (28)(x^6)(4) + (56)(x^5)(8)$
$\quad + (70)(x^4)(16) + (56)(x^3)(32) + (28)(x^2)(64)$
$\quad + (8)(x)(128) + (1)(1)(256)$

$= x^8 + 16x^7 + 112x^6 + 448x^5 + 1120x^4 + 1792x^3$
$\quad + 1792x^2 + 1024x + 256$

B

Evaluate:

17. $\dfrac{18!}{16!2!} = \dfrac{18 \cdot 17 \cdot 16!}{16!2!} = \dfrac{18 \cdot 17}{2 \cdot 1} = 153$

21. $\left[3x - \dfrac{y}{3}\right]^4$

    $a = 3x, \; b = \left(-\dfrac{y}{3}\right), \; n = 4$ Use the formula to expand.

$\dfrac{4!}{4!0!}(3x)^4\left(-\dfrac{y}{3}\right)^0 + \dfrac{4!}{3!1!}(3x)^3\left(-\dfrac{y}{3}\right)^1$

$+ \dfrac{4!}{2!2!}(3x)^2\left(-\dfrac{y}{3}\right)^2 + \dfrac{4!}{1!3!}(3x)^1\left(-\dfrac{y}{3}\right)^3$

$+ \dfrac{4!}{0!4!}(3x)^0\left(-\dfrac{y}{3}\right)^4$

$= 1(81x^4)(1) + 4(27x^3)\left(-\dfrac{y}{3}\right) + 6(9x^2)\left[\dfrac{y^2}{9}\right]$

$\quad + 4(3x)\left[-\dfrac{y^3}{27}\right] + 1(1)\left[\dfrac{y^4}{81}\right]$

$= 81x^4 - 36x^3y + 6x^2y^2 - \dfrac{4}{9}xy^3 + \dfrac{1}{81}y^4$

25. $(2x - 3)^5$
    $a = 2x, b = -3, n = 5$

$$\frac{5!}{5!0!}(2x)^5(-3)^0 + \frac{5!}{4!1!}(2x)^4(-3)^1 + \frac{5!}{3!2!}(2x)^3(-3)^2$$

$$+ \frac{5!}{2!3!}(2x)^2(-3)^3 + \frac{5!}{1!4!}(2x)^1(-3)^4$$

$$+ \frac{5!}{0!5!}(2x)^0(-3)^5$$

$$= 1(32x^5)(1) + 5(16x^4)(-3) + 10(8x^3)(9) + 10(4x^2)(-27)$$

$$+ 5(2x)(81) + 1(1)(-243)$$

$$= 32x^5 - 240x^4 + 720x^3 - 1080x^2 + 810x - 243$$

Find the specified term of the expanded form:

29. $(x + 1)^{10}$, seventh term
    $a = x, b = 1, n = 10, k = 6$ Identify a, b, n, and k.
                                         Note that $k = 7 - 1 = 6$.

$$\frac{10!}{4!6!}(x)^4(1)^6 = 210x^4 \qquad \text{Substitute in the formula.}$$

C

Write the first four terms of the expanded form:

33. $(2x + y)^{24}$
    $a = 2x, b = y, n = 24$ Identify a, b, and n.

$$\frac{24!}{24!0!}(2x)^{24}(y)^0 + \frac{24!}{23!1!}(2x)^{23}(y)^1 \quad \text{Use the formula to expand.}$$

$$+ \frac{24!}{22!2!}(2x)^{22}(y)^2 + \frac{24!}{21!3!}(2x)^{21}(y)^3$$

$$= (1)(2x)^{24}(1) + (24)(2x)^{23}(y) + 276(2x)^{22}(y^2) + 2024(2x)^{21}(y^3)$$

$$= (2x)^{24} + 24(2x)^{23}y + 276(2x)^{22}y^2 + 2024(2x)^{21}y^3$$

37. $(x^2 - 2y^3)^{12}$
    $a = x^2$, $b = -2y^3$, $n = 12$

$$\frac{12!}{12!0!}(x^2)^{12}(-2y^3)^0 + \frac{12!}{11!1!}(x^2)^{11}(-2y\ )$$

$$+ \frac{12!}{10!2!}(x^2)^{10}(-2y^3)^2 + \frac{12!}{9!3!}(x^2)^9(-2y^2)^3$$

$$= 1(x^{24})(1) + 12(x^{22})(-2y^3) + 66(x^{20})(4y^6) + 220x^{18}(-8y^9) + \ldots$$

$$= x^{24} - 24x^{22}y^3 + 264x^{20}y^6 - 1760x^{18}y^9$$

Find the specified term of the expanded form:

41. $(x + 1)^{22}$, eleventh term
    $a = x$, $b = 1$, $n = 22$, $k = 11 - 1 = 10$

$$\frac{22!}{12!10!}(x)^{12}(1)^{10} = 646646x^{12}$$

45. $\left[\frac{x}{2} + 2\right]^{15}$, eighth term
    $a = \frac{x}{2}$, $b = 2$, $n = 15$, $k = 8 - 1 = 7$

$$\frac{15!}{8!7!}\left[\frac{x}{2}\right]^8(2)^7 = 6435\left[\frac{x^8}{256}\right](128)$$

$$= \frac{6435}{2}x^8$$

D
49. If a coin is tossed eight times, the number of ways that exactly four heads and four tails will show is the coefficient of the fifth term of the expansion of $(H + T)^8$. Find the number of ways.

$a = H$, $b = T$, $n = 8$, $k = 5 - 1 = 4$

$$\frac{8!}{4!4!} = \frac{8 \cdot 7 \cdot 6 \cdot 5 \cdot 4!}{4!4!}$$

The variables are not needed, since the coefficient of the term is the answer to the question.

$$= \frac{8 \cdot 7 \cdot 6 \cdot 5}{4 \cdot 3 \cdot 2 \cdot 1}$$

$$= 70$$

Answer:
There are 70 ways you can get 4 heads and 4 tails.

E Maintain Your Skills

Find the logarithm to four decimal places:

53. $\ln 0.0378$

ENTER			DISPLAY	
$\boxed{0.0378}$	$\boxed{\ln x}$		-3.2754462	By calculator.

$\ln 0.0378 \approx -3.2754$

Find the antilogarithm to four decimal places:

57. $\ln x = 2.7586$

ENTER			DISPLAY	
$\boxed{2.7586}$	$\boxed{\text{INV}}$	$\boxed{\ln x}$	15.777739	By calculator.

antiln $2.7586 \approx 15.7777$.